D0204040

GRAPHS AND THEIR USES

NEW MATHEMATICAL LIBRARY

PUBLISHED BY

THE MATHEMATICAL ASSOCIATION OF AMERICA

The New Mathematical Library (NML) was begun in 1961 by the School Mathematics Study Group to make available to high school students short expository books on various topics not usually covered in the high school syllabus. In a decade the NML matured into a steadily growing series of some twenty titles of interest not only to the originally intended audience, but to college students and teachers at all levels. Previously published by Random House and L. W. Singer, the NML became a publication series of the Mathematical Association of America (MAA) in 1975. Under the auspices of the MAA the NML will continue to grow and will remain dedicated to its original and expanded purposes.

GRAPHS AND THEIR USES

by

Oystein Ore

Yale University

Revised and updated edition prepared by

Robin J. Wilson

The Open University, England

34

THE MATHEMATICAL ASSOCIATION OF AMERICA

Library of Congress Catalog Card Number: 90-061132
Complete Set ISBN 0-88385-600-X
Vol. 34 0-88385-635-2

Manufactured in the United States of America

Note to the Reader

This book is one of a series written by professional mathematicians in order to make some important mathematical ideas interesting and understandable to a large audience of high school students and laymen. Most of the volumes in the *New Mathematical Library* cover topics not usually included in the high school curriculum; they vary in difficulty, and, even within a single book, some parts require a greater degree of concentration than others. Thus, while you need little technical knowledge to understand most of these books, you will have to make an intellectual effort.

If you have so far encountered mathematics only in classroom work, you should keep in mind that a book on mathematics cannot be read quickly. Nor must you expect to understand all parts of the book on first reading. You should feel free to skip complicated parts and return to them later; often an argument will be clarified by a subsequent remark. On the other hand, sections containing thoroughly familiar material may be read very quickly.

The best way to learn mathematics is to *do* mathematics, and each book includes problems, some of which may require considerable thought. You are urged to acquire the habit of reading with paper and pencil in hand; in this way, mathematics will become increasingly meaningful to you.

The authors and editorial committee are interested in reactions to the books in this series and hope that you will write to: Anneli Lax, Editor, New Mathematical Library, NEW YORK UNIVERSITY, THE COURANT INSTITUTE OF MATHEMATICAL SCIENCES, 251 Mercer Street, New York, N. Y. 10012.

The Editors

NEW MATHEMATICAL LIBRARY

1 Numbers: Rational and Irrational *by Ivan Niven*
2 What is Calculus About? *by W. W. Sawyer*
3 An Introduction to Inequalities *by E. F. Beckenbach and R. Bellman*
4 Geometric Inequalities *by N. D. Kazarinoff*
5 The Contest Problem Book I Annual High School Mathematics Examinations 1950–1960. Compiled and with solutions *by Charles T. Salkind*
6 The Lore of Large Numbers, *by P. J. Davis*
7 Uses of Infinity *by Leo Zippin*
8 Geometric Transformations I *by I. M. Yaglom, translated by A. Shields*
9 Continued Fractions *by Carl D. Olds*
10 Graphs and Their Uses *by Oystein Ore*
11 } Hungarian Problem Books I and II, Based on the Eötvös
12 } Competitions 1894–1905 and 1906–1928, *translated by E. Rapaport*
13 Episodes from the Early History of Mathematics *by A. Aaboe*
14 Groups and Their Graphs *by I. Grossman and W. Magnus*
15 The Mathematics of Choice *by Ivan Niven*
16 From Pythagoras to Einstein *by K. O. Friedrichs*
17 The Contest Problem Book II Annual High School Mathematics Examinations 1961–1965. Compiled and with solutions *by Charles T. Salkind*
18 First Concepts of Topology *by W. G. Chinn and N. E. Steenrod*
19 Geometry Revisited *by H. S. M. Coxeter and S. L. Greitzer*
20 Invitation to Number Theory *by Oystein Ore*
21 Geometric Transformations II *by I. M. Yaglom, translated by A. Shields*
22 Elementary Cryptanalysis—A Mathematical Approach *by A. Sinkov*
23 Ingenuity in Mathematics *by Ross Honsberger*
24 Geometric Transformations III *by I. M. Yaglom, translated by A. Shenitzer*
25 The Contest Problem Book III Annual High School Mathematics Examinations 1966–1972. Compiled and with solutions *by C. T. Salkind and J. M. Earl*
26 Mathematical Methods in Science *by George Pólya*
27 International Mathematical Olympiads—1959–1977. Compiled and with solutions *by S. L. Greitzer*
28 The Mathematics of Games and Gambling *by Edward W. Packel*
29 The Contest Problem Book IV Annual High School Mathematics Examinations 1973–1982. Compiled and with solutions *by R. A. Artino, A. M. Gaglione and N. Shell*
30 The Role of Mathematics in Science *by M. M. Schiffer and L. Bowden*
31 International Mathematical Olympiads 1978–1985 and forty supplementary problems. Compiled and with solutions by *Murray S. Klamkin.*
32 Riddles of the Sphinx *by Martin Gardner*
33 U.S.A. Mathematical Olympiads 1972–1986. Compiled and with solutions by *Murray S. Klamkin.*
34 Graphs and Their Uses, *by Oystein Ore*. Revised and updated *by Robin J. Wilson.*

Other titles in preparation.

Contents

INTRODUCTION TO THE FIRST EDITION

The term "graph" in this book denotes something quite different from the graphs you may be familiar with from analytic geometry or function theory. The kind of graph you probably have dealt with consisted of the set of all points in the plane whose coordinates (x, y), in some coordinate system, satisfy an equation in x and y. The graphs we are about to study in this book are simple geometrical figures consisting of points and lines connecting some of these points; they are sometimes called "linear graphs". It is unfortunate that two different concepts bear the same name, but this terminology is now so well established that it would be difficult to change. Similar ambiguities in the names of things appear in other mathematical fields, and unless there is danger of serious confusion, mathematicians are reluctant to alter the terminology.

The first paper on graph theory was written by the famous Swiss mathematician Euler, and appeared in 1736. From a mathematical point of view, the theory of graphs seemed rather insignificant in the beginning, since it dealt largely with entertaining puzzles. But recent developments in mathematics, and particularly in its applications, have given a strong impetus to graph theory. Already in the nineteenth century, graphs were used in such fields as electrical circuitry and molecular diagrams. At present there are topics in pure mathematics—for instance, the theory of mathematical relations—where graph theory is a natural tool, but there are also numerous other uses in connection with highly practical questions: matchings, transportation problems, the flow in pipeline networks, and so-called "programming" in general. Graph theory now makes its appearance in such diverse fields as economics, psychology and biology. To a small extent puzzles remain a part of graph theory, particularly if one includes among them the famous *four color map problem* that intrigues mathematicians today as much as ever.

In mathematics, graph theory is classified as a branch of topology; but it is also strongly related to algebra and matrix theory.

In the following discussion we have been compelled to treat only the simplest problems from graph theory; we have selected these with the intention of giving an impression, on the one hand, of the kind of analyses that can be made by means of graphs and, on the other hand, of some of the problems that can be attacked by such methods. Fortunately, no great apparatus of mathematical computation needs to be introduced.

O. Ore

INTRODUCTION TO THE REVISED EDITION

In preparing this edition I have endeavored to stick as closely as possible to Oystein Ore's original intentions. However, I have felt free to make a large number of minor changes in the presentation and layout of the material, and I have updated the terminology and notation so as to bring them in line with contemporary usage. I have also added new material on interval graphs, the travelling salesman problem, bracing frameworks, shortest route problems, and coloring maps on surfaces. Most of the diagrams in the book have been redrawn.

I have always regarded Ore's text as a classic, and working on this second edition has served to reinforce this view. It is my hope that this edition will enable a new generation of readers to derive as much pleasure from Ore's book as my generation did in the 1960s and 1970s.

R. J. Wilson

CHAPTER ONE

What is a Graph?

1.1 Team Competitions

Suppose that your school football team belongs to a league in which it plays the teams of certain other schools. Call your own team a and the other teams b, c, d, e and f, and assume that there are 6 teams altogether. After a few weeks of the season have passed some of the teams will have played each other—for instance,

a has played c, d, f
b has played c, e, f
c has played a, b
d has played a, e, f
e has played b, d, f
f has played a, b, d, e.

To illustrate this situation we can use a geometric diagram. Each team can be represented by a point or a little circle, and two such points can be connected by a straight line whenever the teams they represent have played their game. Then the above list of completed games can be presented as in Figure 1.1.

A figure such as the one drawn in Figure 1.1 is called a *graph*. It consists of certain points a, b, c, d, e, f, called its *vertices*, and certain line segments connecting vertices (such as ac, eb, etc.), called the *edges* of the graph. We shall call this graph G.

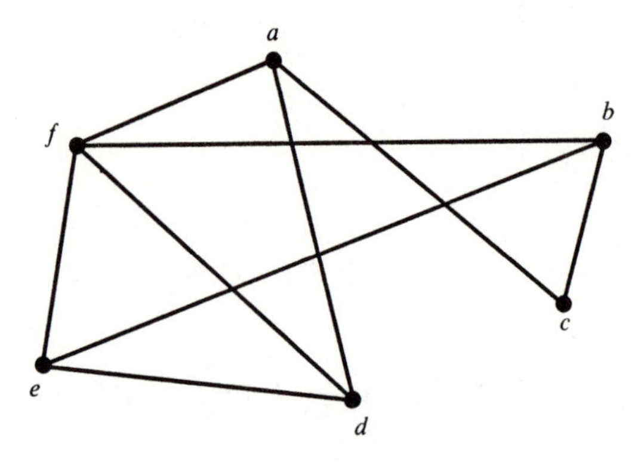

Figure 1.1

It may happen, as we see from Figure 1.1, that the edges of a graph intersect without the intersection being a vertex; this complication is due to the fact that we drew our graph in the plane. Therefore, it might have been more appropriate to represent the edges as threads passing over each other in space; but in any case the marking of the vertices should be done with sufficient care to prevent confusion.

Any set of games played in a team tournament can be depicted as a graph in the manner described. On the other hand, if one has some graph (that is, a figure consisting of points or vertices connected by line segments or edges), then it can be interpreted as the diagram of such a competition. As an illustration let us take the graph drawn in Figure 1.2. The figure may be considered to depict a competition between 8 teams; a has played with the teams b, e, d, while b has played with a, f, g, c, and so on.

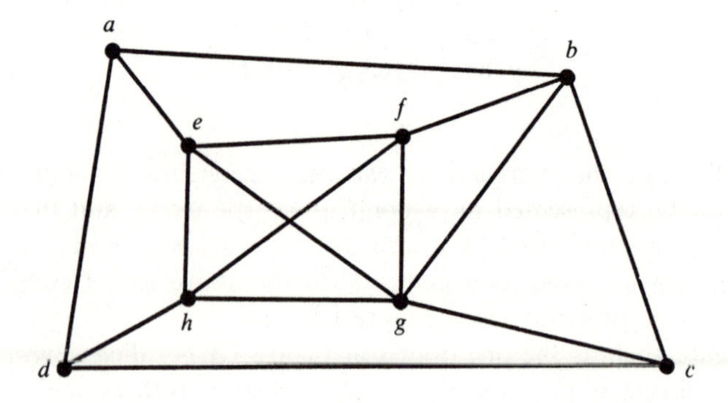

Figure 1.2

Problem Set 1.1

1. Draw the graph of the games played at mid-season in your football or baseball circuit.
2. Write a complete list of the games played in the graph in Figure 1.2.
3. How many vertices and edges are there in the graphs in Figure 1.1 and Figure 1.2, respectively?

1.2 Null Graphs and Complete Graphs

There are certain special graphs which turn up in many uses of graph theory. For the moment let us stick to our interpretation of a graph as a pictorial record of team competitions. Before the season starts, when no games have been played, there will be no edges in the graph. Thus the graph will consist only of *isolated vertices*—that is, vertices at which there are no edges. We call a graph of this kind a *null graph*. In Figure 1.3 we have drawn such graphs for 1, 2, 3, 4 and 5 teams or vertices. These null graphs are commonly denoted by the symbols N_1, N_2, N_3, and so on, so that in general N_n is the null graph with n vertices and no edges.

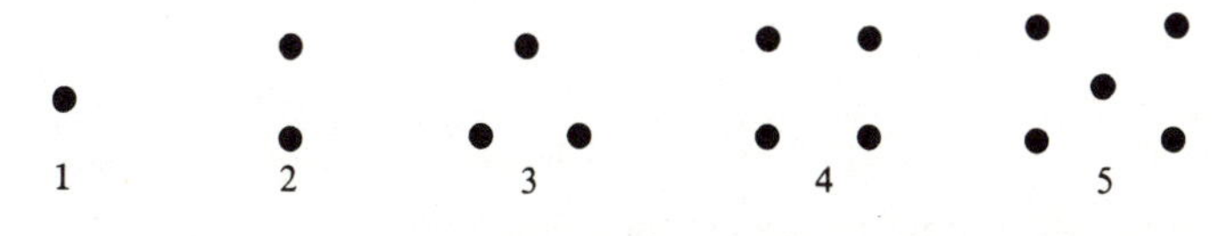

Figure 1.3

Next let us go to another extreme. When the season is over we suppose that each team has played just once with every other team. Then in the game graph each pair of vertices is connected by an edge. Such a graph is called a *complete graph*. Figure 1.4 shows the complete graphs for $n = 1, 2, 3, 4, 5$ vertices. We denote these complete graphs by K_1, K_2, K_3, K_4, K_5, respectively, so that in general K_n consists of n vertices and the edges connecting all pairs of these vertices. It can be drawn as a polygon with n sides and with all its diagonals.

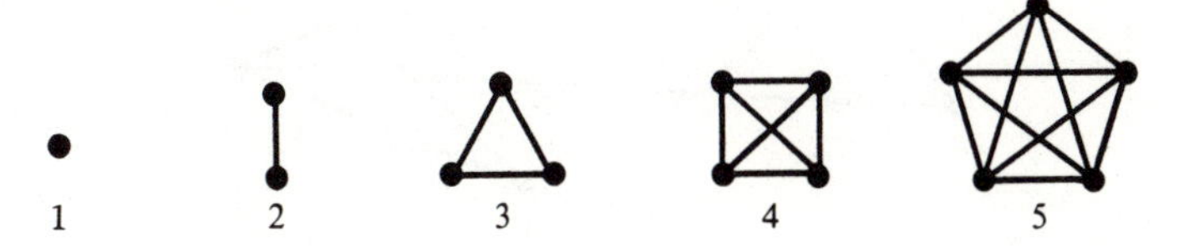

Figure 1.4

When one has drawn some graph—for instance, the graph G in Figure 1.1—one can always make it into a complete graph with the same vertices by adding the missing edges—that is, the edges which correspond to games still to be played. In Figure 1.5 we have done this for the graph G in Figure 1.1. (Games not yet played are represented by broken lines.)

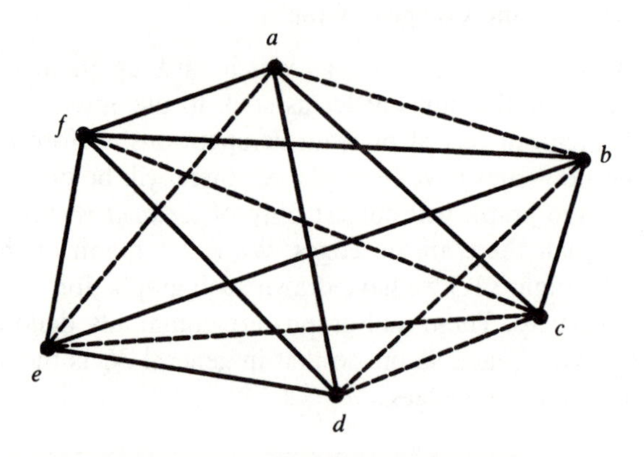

Figure 1.5

One can also draw separately the graph consisting exclusively of the unplayed, future games. In the case of the graph G, this results in the graph depicted in Figure 1.6.

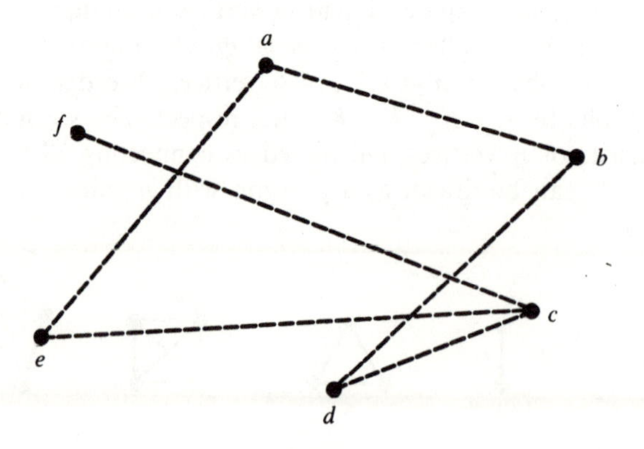

Figure 1.6

This new graph in Figure 1.6 we call the *complement* of the graph G in Figure 1.1, and it is customary to denote it by $\overline{G}$. If we take the complement of $\overline{G}$, we get back to G; together the edges in the two graphs G and $\overline{G}$ make up the complete graph connecting their vertices.

Problem Set 1.2

1. Draw the complement of the graph in Figure 1.2.
2. How many edges have the complete graphs K_5, K_6 and K_7?
3. Express in terms of n the number of edges in the complete graph K_n.

1.3 Isomorphic Graphs

Notice that, in drawing the graph in Figure 1.1, we have a good deal of freedom.

First, there is no necessity for the edges to be straight lines. Any kind of curves will do as long as they connect the same vertices as before. For example, we can present the graph in Figure 1.1 in the following form (Figure 1.7):

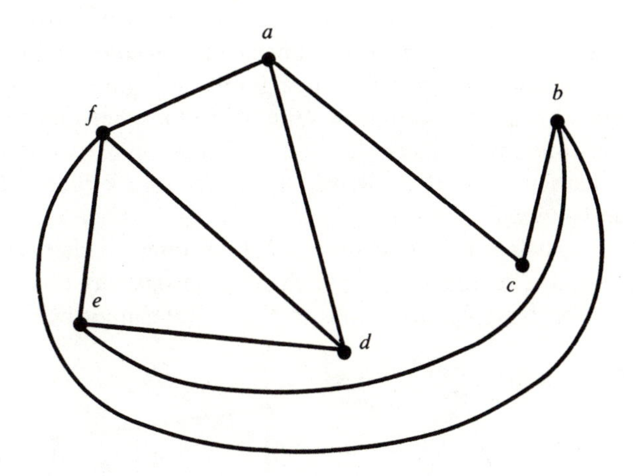

Figure 1.7

Second, we can place the vertices in arbitrary positions in the plane. The graph in Figure 1.1, for instance, can be drawn with the vertices placed as in Figure 1.8.

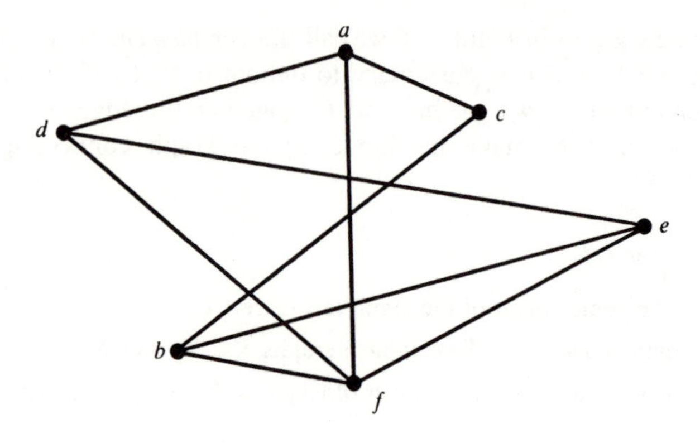

Figure 1.8

If we consider the three graphs in Figure 1.1, Figure 1.7 and Figure 1.8 as the graphs of tournament games, they all contain exactly the same information in regard to which teams have played each other—that is, they are in a sense the same graph. This leads us to say in general that two graphs, G_1 and G_2 are *isomorphic* if they represent the same situation. In other words, if G_1 and G_2 are isomorphic, then they have the same number of vertices, and whenever two vertices in G_1, say b_1 and c_1, are connected by an edge, then the corresponding vertices b_2 and c_2 in G_2 are also connected by an edge, and vice versa. According to this definition the three graphs in Figure 1.1, Figure 1.7, and Figure 1.8 are isomorphic in spite of the fact that they have been drawn in different manners. (The term "isomorphic" is a much used one in mathematics; it is derived from the Greek words *iso*—the same, and *morphe*—form.)

Often one is faced with the problem of deciding whether two graphs are isomorphic. At times there are obvious reasons why this cannot be the case. For example, the graphs in Figure 1.9 cannot be isomorphic

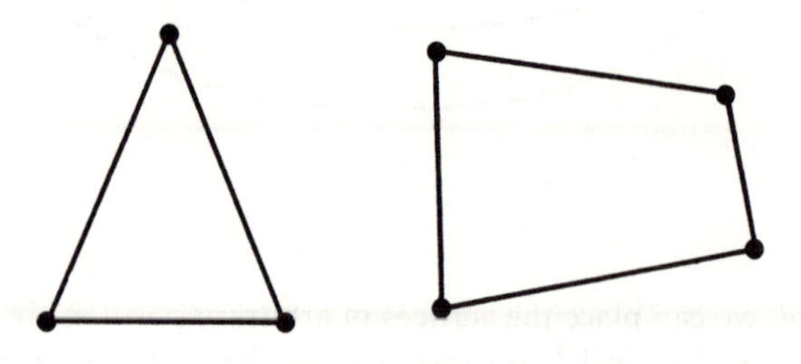

Figure 1.9

because they do not have the same number of vertices. Nor can the graphs in Figure 1.10 be isomorphic, since they do not have the same number of edges.

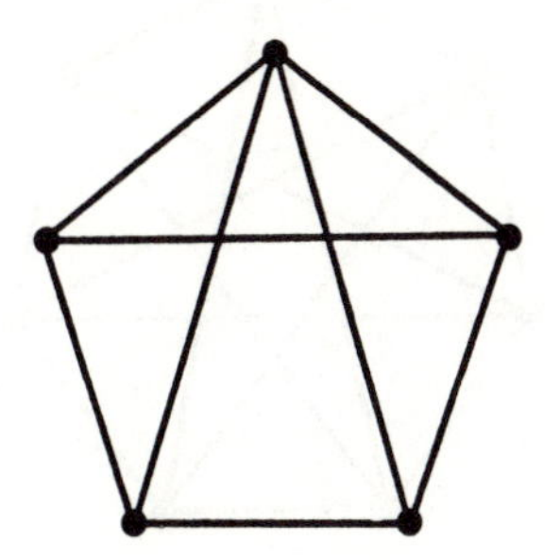
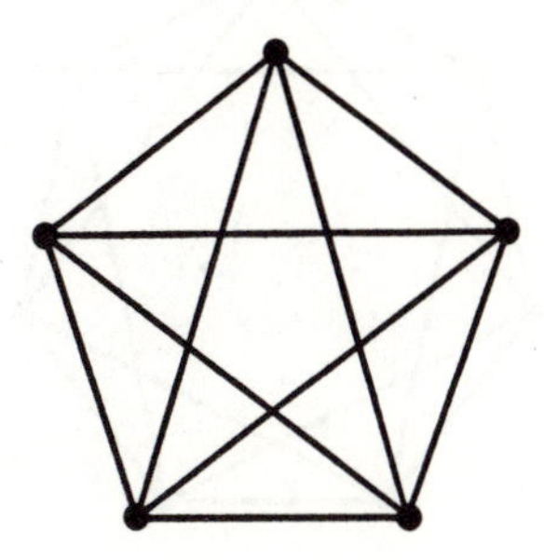

Figure 1.10

Slightly more subtle reasoning is required to show that the two graphs in Figure 1.11 are not isomorphic. One can observe, however, that in the first graph the vertices a, b, c, d with two edges emerging from them are joined in pairs (ab and cd), whereas in the second graph they are not so joined. In other words, no matter how we name the vertices of the second graph, we shall not be able to match pairs of vertices connected by an edge in one graph with corresponding pairs of vertices connected by an edge in the other graph.

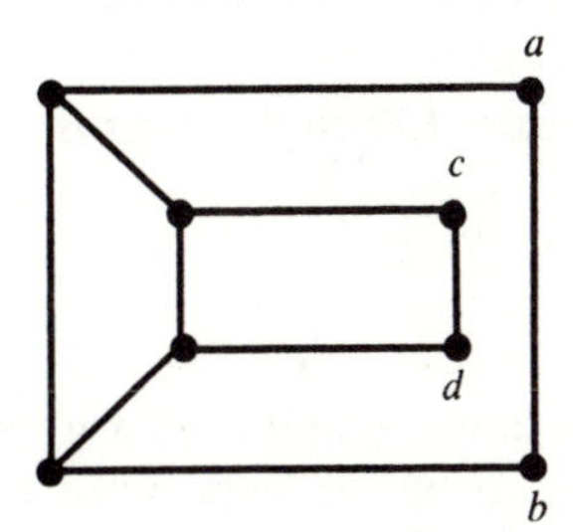

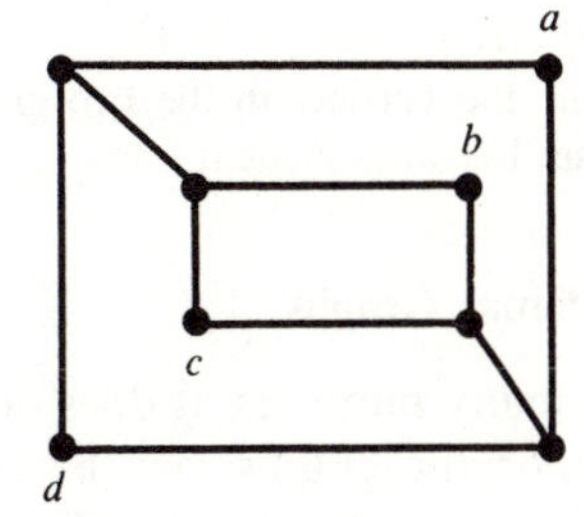

Figure 1.11

When there is no obvious way to show that two given graphs are not isomorphic, it may be quite difficult to decide whether we can name the vertices in such a manner that we obtain an isomorphism between the graphs. As an example, consider the two graphs in Figure

1.12; they are actually isomorphic, as we ask you to show in Problem 3, below.

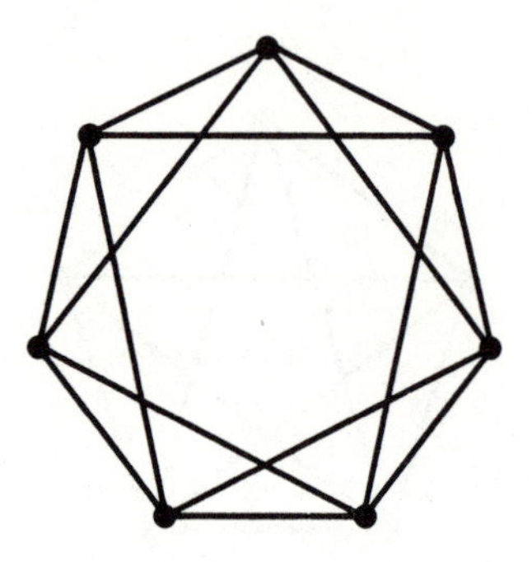
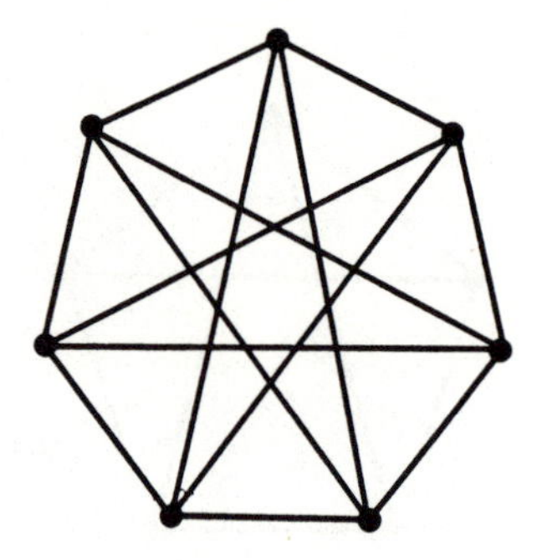

Figure 1.12

The determination of efficient criteria for deciding whether two given graphs are isomorphic is one of the major concerns in current graph theory. As the number of vertices increases, the number of ways of naming them grows very fast, and isomorphisms between the graphs become very hard to find, even with a computer.

Problem Set 1.3

1. Show that the graphs in Figure 1.1, Figure 1.2 and Figure 1.6 are not isomorphic to each other.
2. Give another reason why the two graphs in Figure 1.11 cannot be isomorphic.
3. Name the vertices in the two graphs in Figure 1.12 so that their isomorphism becomes evident.

1.4 Planar Graphs

For many purposes, it does not matter how a graph is drawn; that is, isomorphic graphs may be considered to be the same since they give the same information. This was certainly the case in our initial interpretation of graphs as the record of games between teams. However, as we shall point out presently, there are purposes for which it is essential that a graph can be drawn in a particular way. Let us compare the two isomorphic graphs in Figure 1.1 and Figure 1.7. In the first drawing, the edges intersect at 5 points that are not vertices of the graph. On the other hand, in Figure 1.7 the edges intersect only at vertices.

A graph which can be drawn in such a way that the edges have no intersections or common points except at the vertices is called a *planar graph*. Thus the graph in Figure 1.1 is planar because there exists a representation of it in the plane as in Figure 1.7.

A planar graph can be interpreted as a road map showing the connections between various road junctions or villages. For instance, the map in Figure 1.13 indicates that there are 7 junctions, a to g, some of which are directly connected by roads, such as ag, bc, fe, and so on. Conversely, a road map can be considered to be a planar graph. Similarly, a city map can be regarded as a planar graph with the streets as edges and the street intersections as vertices—see Figure 1.14.

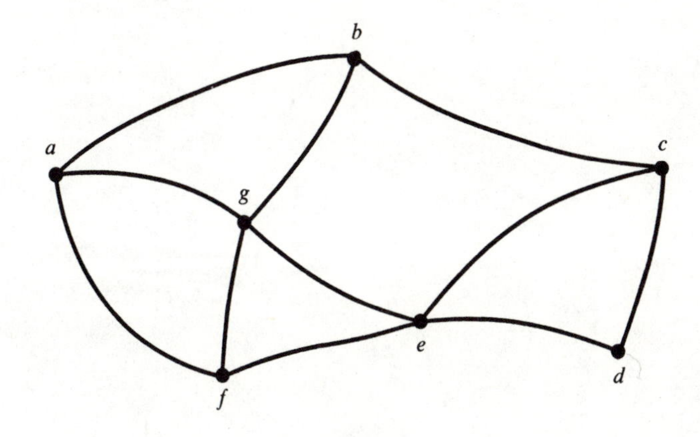

Figure 1.13

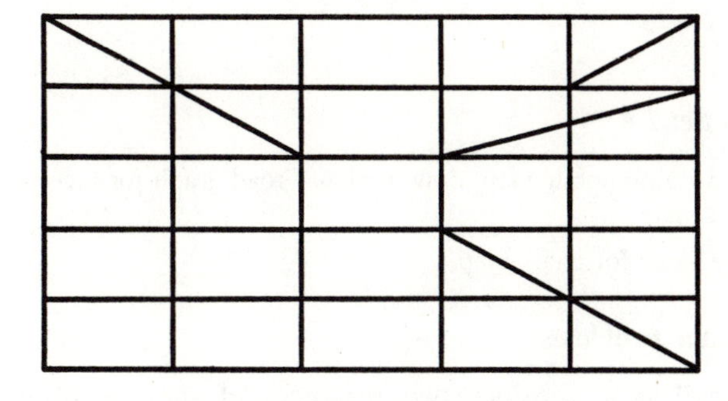

Figure 1.14

Modern technology has changed many things, and we must recognize that it has modified the preceding simple conception of road maps as planar graphs. To our road net have been added throughways with limited access, so that often two roads cross without permitting passage from one to the other; in other words, the edges of the map graph intersect at points which are not road junctions and thus the corresponding graph is not a planar graph. Figure 1.15 shows a road intersection which illustrates this very effectively.

Figure 1.15

Problem Set 1.4

1. From an automobile map, draw a planar road graph for a certain section of your state.
2. Do the same for a city map.

1.5 Planar Problems

We shall now consider two instances of the use of graphs in problem solving. In both cases it is essential to decide whether or not a graph can be drawn in the plane without intersections of the edges.

As our first illustration, let us turn to a very ancient puzzle (sometimes called the *Utilities Problem*):

Three houses have been built on a piece of land, and three wells have been dug for the use of the occupants. The nature of the land and the climate are such that one or another of the wells frequently runs dry; it is therefore important that the people of each house have access to each of the three wells. After a while, the residents a, b and c develop rather strong dislikes for one another and decide to construct paths to the three wells x, y, z in such a manner that they avoid meeting each other on their way to and from the wells.

In Figure 1.16, we see the graph of the arrangement in which each owner uses the most direct paths to the wells. These paths or edges intersect in many points aside from the houses a, b, c, and the wells x, y, z. The number of intersections can be reduced to a single one, provided we draw the paths as indicated in Figure 1.17.

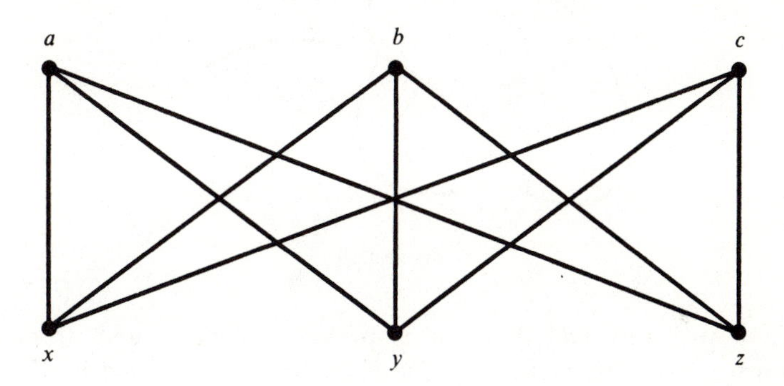

Figure 1.16

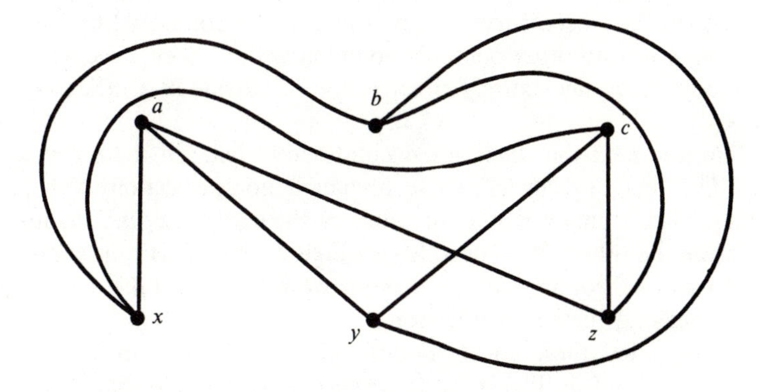

Figure 1.17

The question we should like to answer is the following: can we trace the paths so that the graph is planar—that is, without *any* edge intersections? Try as you may, you will find no such tracing. However, our inability to solve this problem by trial and error does not constitute a mathematical proof that no such tracing exists. A mathematical proof can be given and is based on the following result (see Figure 1.18):

JORDAN CURVE THEOREM. *Suppose C is a continuous closed curve in the plane—it may be a polygon, a circle, an ellipse, or some more complicated type of curve. Then C divides the plane into an outer part and an inner part, so that whenever any point p in the inner part is connected to a point q in the outer part by a continuous curve L in the plane, then L intersects C.*

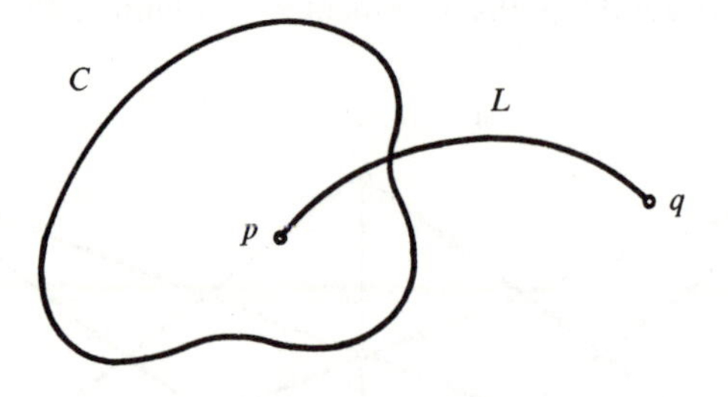

Figure 1.18

You probably feel that this is perfectly obvious, and, from an intuitive geometric view, it is. The difficulty lies in the precise definition of "curve", which we omit here, together with the proof of the Jordan curve theorem. You may take the theorem as an evident fact.

This theorem implies the intuitively obvious result that if any two points on the closed curve C, say a and y, are connected by a curve ay which has no other points in common with C, then, except for its end points, ay lies entirely either inside or outside of C. (See Figure 1.19.)

Suppose next that there are 4 points on C lying in the order $abyz$ and that there are curves ay and bz having no intersections with each other. This is only possible when one of the curves, say ay, lies inside C while the other, bz, is outside (see Figure 1.19). This can be proved by Jordan's theorem, but you may (as we have done) take it as a fact that needs no further justification.

Finally, let there be 6 points on C following in the order a, x, b, y, c, z (see Figure 1.19). Then it is impossible that there are

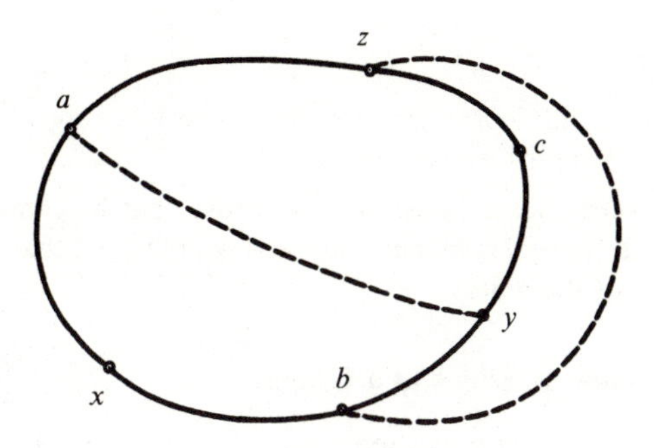

Figure 1.19

three connecting curves *ay*, *bz* and *cx* without intersections. To see this, observe that the three curves must be placed into two regions, the inside of C and the outside of C; therefore at least two of the curves fall into the same region and, by the considerations above, this would lead to intersections.

This argument applies immediately to our problem of the three unfriendly neighbors and their wells. Suppose that the corresponding graph in Figure 1.16 were planar. Then, in any drawing without edge intersections, the edges *ax*, *xb*, *by*, *yc*, *cz* and *za* would form a closed curve in the plane. But then, for the reason we just explained, there can be no edges *ay*, *bz* and *cx* without intersections.

This illustration of the use of planar graphs may seem somewhat trivial; however, one should never despise these apparently small, but puzzling, problems. In numerous instances, they have been the seeds from which important mathematical ideas have evolved. It may remind us also that a heavy machinery of symbols and formulas is not always the best criterion for judging the depth of a mathematical theory.

We can also indicate an application of planar graphs to an eminently practical problem. In addition to the previous interpretations, a graph can be thought of as the diagram for an electrical network, with the edges representing the conducting wires connecting the various junctions. One of the most effective ways of mass producing a standard network for a radio or television set is to 'print' the wires on a base of board or plastic. In order for this to be feasible, the network graph in question must have a planar representation, since otherwise the intersection of two edges would produce a short circuit in the system. Such applications of planar graphs have become increasingly important in recent years with the rapid developments in electronics.

Problem Set 1.5

1. Show how each of four neighbors can connect his house with the other three houses by paths which do not cross.
2. A fifth person builds a house nearby; prove that he cannot connect his house with all the others by non-intersecting paths, but that he can connect it with three of the others.

1.6 The Number of Edges in a Graph

In introducing a graph as the record of a series of played games, we assumed that at most one game was played between any two teams. It may of course happen that two teams play many games, as they do in the baseball leagues. We can take this into account in the graph by drawing several edges ab connecting the two corresponding teams or vertices (Figure 1.20). We then say that the graph has *multiple edges*.

Figure 1.20

Instead of actually drawing the various edges between a and b, we could also use a single edge and assign a number or *multiplicity* to it to indicate how many times this edge should be repeated (Figure 1.21). On a road map, it is of course customary to draw each road separately between the two junctions.

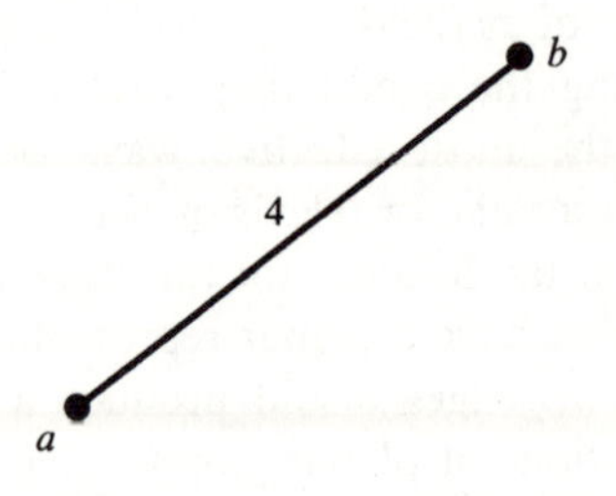

Figure 1.21

At each non-isolated vertex a in a graph G, there will be some edges having a as an endpoint; these edges are said to be *incident* to a. The number of such edges we usually denote by $\deg(a)$, and call it the *degree* of a. To illustrate, we observe that the graph in Figure 1.1 has degrees

$$\deg(a) = \deg(b) = \deg(d) = \deg(e) = 3,$$
$$\deg(f) = 4, \qquad \deg(c) = 2.$$

For many purposes we are interested in finding the number of edges in the graph. They can of course be counted directly, but it is often easier to count the number of edges at each vertex and add them. Then each edge has been counted twice, once at each of its two end points, so the number of edges in the graph is half this sum. For instance, the number of edges in the graph in Figure 1.1 is

$$\tfrac{1}{2}\{\deg(a) + \deg(b) + \deg(c) + \deg(d) + \deg(e) + \deg(f)\} = 9,$$

as we also see directly.

To formulate this quite generally, assume that G is a graph with n vertices $a_1, a_2, \ldots, a_n$, and having as degrees the numbers

$$\deg(a_1), \deg(a_2), \ldots, \deg(a_n).$$

Then the number m of edges in G is, by our argument,

$$m = \tfrac{1}{2}\{\deg(a_1) + \cdots + \deg(a_n)\}.$$

This result is sometimes called the *handshaking lemma*, and has the consequence that *in any graph the sum of the degrees*

$$\deg(a_1) + \cdots + \deg(a_n)$$

is an even number—namely, twice the number of edges. It is due to the Swiss mathematician Leonhard Euler (1707–1783), whom you will meet again in Chapter 2.

In a graph there are two types of vertices, the *odd vertices* whose degree is an odd number, and the *even vertices* whose degree is an even number. In the case of the graph in Figure 1.1, the vertices a, b, d, e are odd while the vertices c and f are even. When the vertices are taken in alphabetical order the sum of the degrees becomes

$$3 + 3 + 2 + 3 + 3 + 4 = 18.$$

This sum is even, for there are 4 terms which are odd numbers.

To decide in general whether a sum of integers is odd or even, we can disregard the even terms; the sum is even or odd depending upon whether it contains an even or an odd number of odd summands. When we apply this observation to the fact that the sum of the degrees is even, we arrive at the following result:

THEOREM 1.1. *A graph has an even number of odd vertices.*

(We include in this statement the case where there are no odd vertices, since 0 is an even number.)

There are special graphs in which all degrees are the same:

$$\deg(a_1) = \cdots = \deg(a_n) = r.$$

The graph is then called *regular of degree r* and, according to the handshaking lemma, the number of its edges is

$$m = \tfrac{1}{2}nr,$$

where n is the number of vertices. The graphs in Figure 1.22 are regular of degree 3 and 4, respectively.

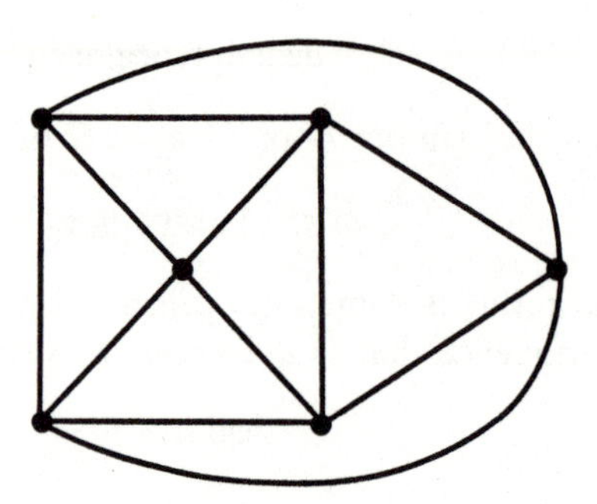

Figure 1.22

In the complete graph K_n with n vertices there are $n - 1$ edges from each vertex to the others, so that K_n is regular of degree $n - 1$. The null graph N_n is also regular in a trivial way, since the degree of each vertex is 0.

Problem Set 1.6

1. Check the handshaking lemma for the number of edges in the graphs in Figures 1.2 and 1.6.
2. Verify that, for these graphs, the number of odd vertices is even.

1.7 Interval Graphs

Consider the following open intervals on a line

$$(0,2), (1,4), (2,5), (3,4), (3,8), (6,9)$$

(see Figure 1.23); recall that the open interval (a, b) is the set of all points lying between a and b.

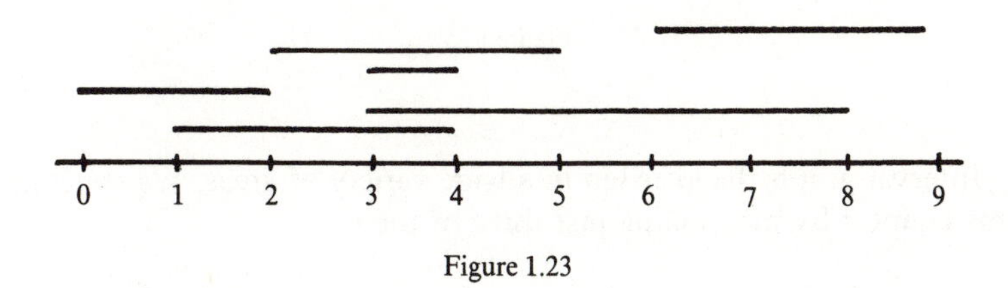

Figure 1.23

We can construct a graph from these intervals by taking the intervals as vertices, and joining two such vertices by an edge whenever the corresponding intervals overlap. For example, the intervals $(0,2)$ and $(1,4)$ overlap, so the corresponding vertices are joined, whereas the intervals $(0,2)$ and $(2,5)$ do not overlap, so the corresponding vertices are not joined. The graph arising from the intervals in Figure 1.23 is shown in Figure 1.24.

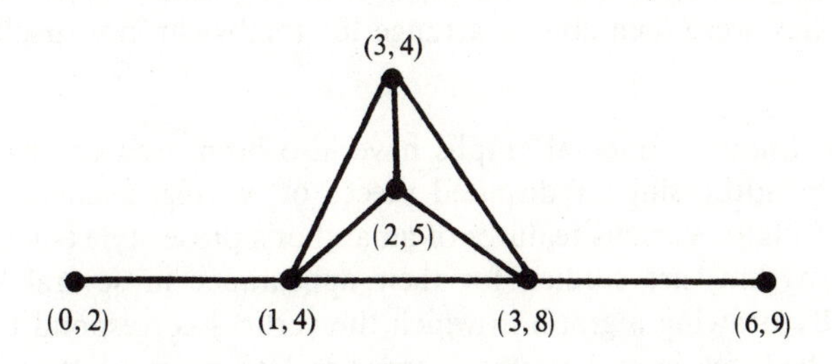

Figure 1.24

Any graph which arises in this way from a set of intervals is called an *interval graph*. For example, the complete graph K_4 is an interval graph, since it arises from the intervals $(1,4)$, $(2,5)$, $(3,4)$ and $(3,8)$—see Figure 1.25.

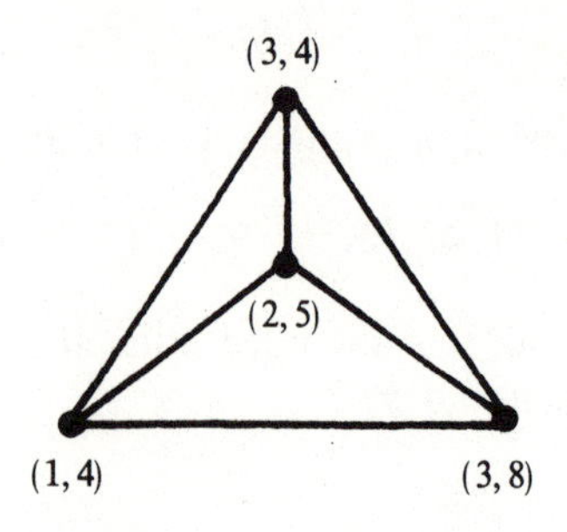

Figure 1.25

Interval graphs have arisen in a wide variety of areas. We conclude this chapter by mentioning just three of them.

Archaeology. Interval graphs have been used by various archaeologists when trying to arrange certain events chronologically. In one experiment a group of archaeologists investigated the artifacts in a large number of tombs in an attempt to arrange these tombs chronologically. Assuming that if two different artifacts occurred together in the same tomb then their time periods must have overlapped, they constructed a graph in which the vertices correspond to the artifacts and the edges correspond to pairs of artifacts which appeared together in some tomb. By representing this graph as an interval graph and interpreting the intervals as time periods during which the artifact was in use, they were then able to arrange the tombs chronologically.

Literary Analysis. Interval graphs have also been used to investigate the likely authorship of disputed pieces of writing, such as certain works of Plato. Various features of an author's prose style (such as the use of rhythm) are studied for their appearance in several literary works. By drawing a graph in which the vertices correspond to these literary features and the edges correspond to pairs of them which occur together in the same work, we obtain a situation very similar to our archaeological example. As before, we can then investigate whether the resulting graph can be represented as an interval graph, and we can thereby attempt to arrange the works in chronological order. By doing this, it has sometimes been possible to relate the style of the disputed piece of writing to that of the author in question, and thereby to determine the likely authorship.

Genetics. Interval graphs arose originally from a problem in genetics —namely, to determine whether the fine structure inside the gene is arranged in a linear manner. In analyzing the genetic structure of a particular virus, the geneticist Seymour Benzer considered the mutations arising when part of the gene is missing. In particular, he was interested in mutations whose missing segments overlap, and he drew a graph in which the vertices correspond to mutations and the edges correspond to pairs of mutations whose missing segments overlap. By representing this graph as an interval graph, he was able to show that (for that virus) the evidence for a linear arrangement inside the gene was overwhelming.

Problem Set 1.7

1. Draw the interval graph of the following set of open intervals:

$$(0,1),(0,5),(1,6),(2,3),(2,7),(4,5),(6,7).$$

2. Show that the graph of Figure 1.1 is not an interval graph.

CHAPTER TWO

Connected Graphs

2.1 Connected Components

Assume again that we have a graph G, not necessarily planar, which we shall now think of as a road map. We may then begin a trip in G at some vertex a, following first an edge or road ab to some junction b, then from b to c on another connecting road bc, and so on. We shall place no restriction on our meandering along the roads; we may pass the same place several times, and even use the same roads over again.

If on this trip we arrive at some vertex t, we say that t is *connected* to a in the graph. This means that there are roads leading from a to t. If we have passed the same locality more than once we can eliminate a circular route and make the trip from a to t more direct. A route in G that passes no vertices twice is called a *path*; the route in Figure 2.1 is a path.

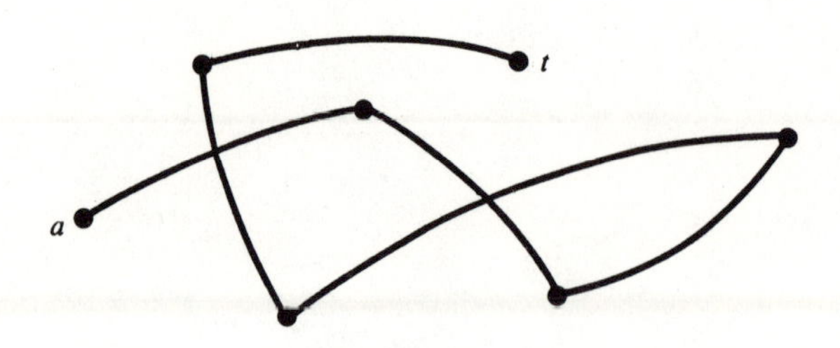

Figure 2.1

A route possibly passing the same vertices several times but never using the same piece of road over again is called a *trail* (Figure 2.2). If the trail returns to the starting point we call it *cyclic* or *circular*, while a returning path is called a *cycle*. Thus a cyclic trail may intersect itself at some of the vertices, but in a cyclic path only the starting vertex is revisited, as the endpoint.

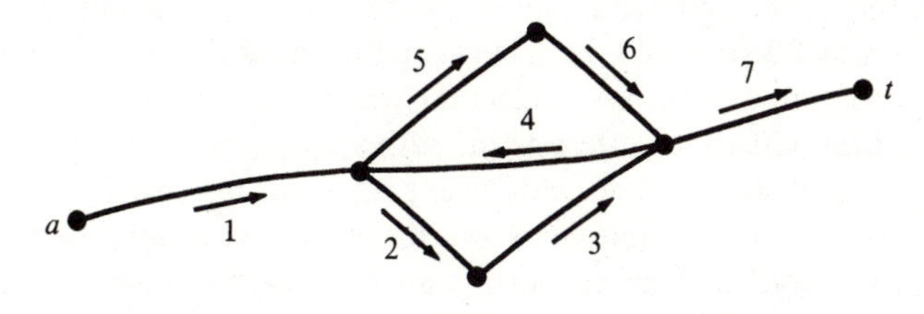

Figure 2.2

Let us illustrate these concepts on the graph in Figure 1.1.

The edge sequence *adfeb* is a path;

the sequence *afdefb* is a trail;

the sequence *afedfbca* is a cyclic trail, while *acbfeda* is a cycle.

When every vertex in a graph is connected to every other vertex by an arc, we say that the graph is *connected*. All graphs we have used as illustrations are connected, except the null graphs. If a graph is not connected, we cannot reach all vertices by arcs from any given vertex a. Those vertices that can be reached by arcs from a vertex a, and the edges incident to them, we call the *connected component* of a. In this manner, the whole graph falls into connected components with no edges or arcs connecting the separate components.

In Figure 2.3, we have illustrated a graph with 4 connected components, one of them an isolated vertex. From the map point of view, we may consider it the road graph of islands, each island having a connected road system. For many considerations in graph theory, we can suppose that the graph is connected, since we can examine separately the properties of each connected component.

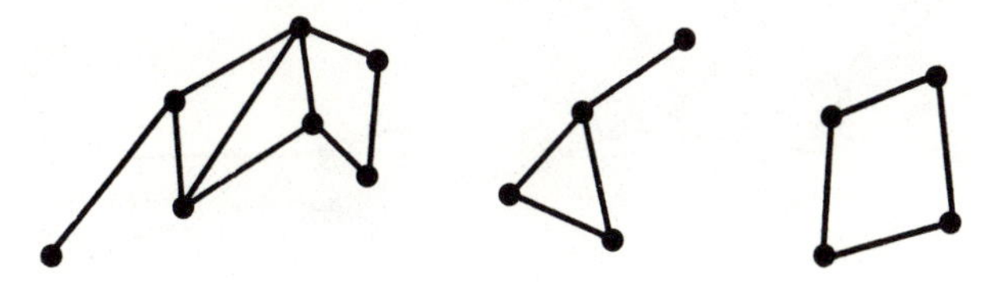

Figure 2.3

2.2 The Problem of the Bridges of Königsberg

The theory of graphs is one of the few fields of mathematics with a definite birth date. The first paper relating to graphs was written by the Swiss mathematician Leonhard Euler (1707–1783), and it appeared in the 1736 volume of the publications of the Academy of Sciences in St. Petersburg (Leningrad). Euler (pronounced 'oyler') is one of the most impressive figures in the history of science. In 1727, when he was 20 years old, he was invited to the Russian academy. He had already studied theology, oriental languages and medicine before he gave free rein to his interests in mathematics, physics, and astronomy. His skill in all these fields was great, and his productivity was enormous. About the time he wrote the paper on graphs he lost his sight in one eye, and as an older man he became totally blind, but even this did not slow the flow of his writings. A considerable time ago Swiss mathematicians, particularly those of his native town of Basel, began an edition of Euler's complete works. When finished, it will contain over 80 volumes.

Euler began his paper on graphs by discussing a puzzle, the so-called *Königsberg Bridges Problem*. The city of Königsberg (now Kaliningrad) in East Prussia is located on the banks and on two islands of the river Pregel. The various parts of the city were connected by seven bridges. On Sundays the burghers would take their promenade around town, as is usual in German cities. The problem arose: *is it possible to plan this "Spaziergang" in such a manner that, starting from home, one can return there after having crossed each river bridge just once*?

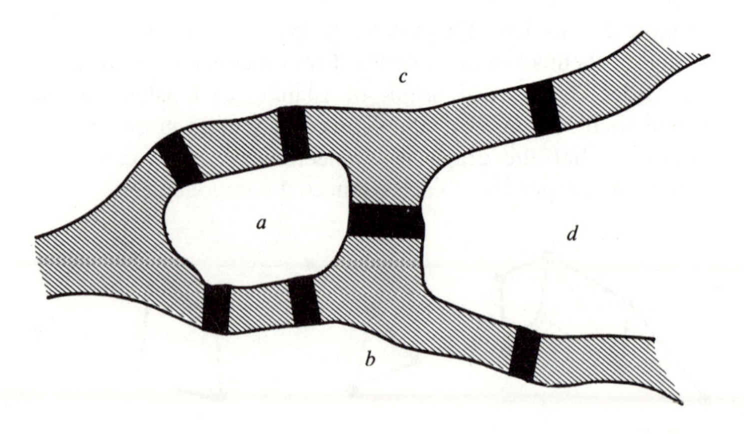

Figure 2.4

A schematic map of Königsberg is reproduced in Figure 2.4. The four parts of the city are denoted by the letters a, b, c and d. Since we are interested only in the bridge crossings we may think of a, b, c, d as the vertices of a graph, with connecting edges corresponding to the bridges. This graph (not used by Euler) is drawn in Figure 2.5.

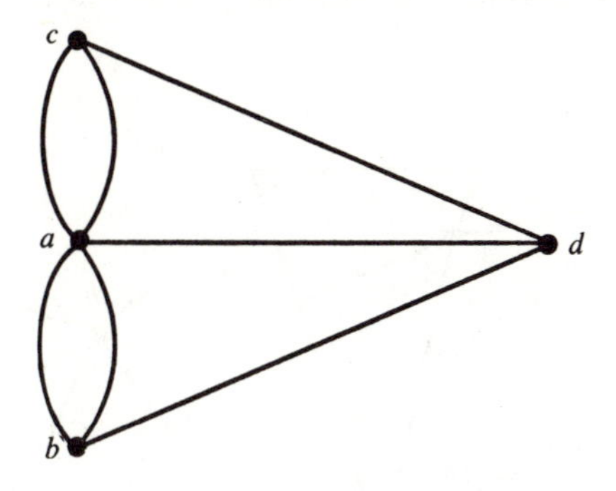

Figure 2.5

Euler's discussion showed that this graph cannot be traversed completely in a single cyclic trail; in other words, no matter at which vertex one begins, one cannot cover the graph and come back to the starting point without retracing one's steps. Such a trail would have to enter each vertex as many times as it departs from it; hence it requires an even number of edges at each vertex, and this condition is not fulfilled in the graph representing the map of Königsberg.

2.3 Eulerian Graphs

After his introduction on the Königsberg bridges, Euler turned to the general problem: *in which graphs is it possible to find a cyclic trail running through all edges just once*? Such a trail is now called an *Eulerian trail*, and a graph with an Eulerian trail is an *Eulerian graph*.

To have an Eulerian trail, the graph must be connected. As in the discussion of the Königsberg Bridges Problem, we see that any Eulerian trail must enter and then exit the same number of times at each vertex—that is, all degrees must be even. Thus two necessary conditions for a graph to contain an Eulerian trail are: connectedness, and evenness of all degrees. Euler proved that these conditions are also sufficient.

THEOREM 2.1. *A connected graph with even degrees has an Eulerian trail.*

PROOF. Suppose that we begin a trail T at some vertex a and continue it as far as possible, always departing from a vertex on an edge which we have not traversed before. This process must stop after a while since we shall run out of new edges. But since there are an even number of edges at each vertex, there is always an exit except at the initial vertex a. Thus T must come to a halt at a (see Figure 2.6).

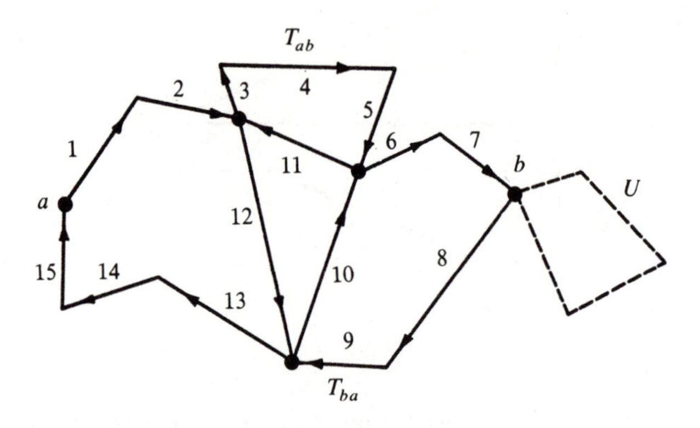

Figure 2.6

If T passes through all edges, we have obtained an Eulerian trail as we wanted. If not, there will be some vertex b lying on T where there are edges not traversed by T. As a matter of fact, since T has an even number of edges at b, there must be an even number of edges at b which do not belong to T, and the same must be true for all vertices where there are untraversed edges.

We now start a trail U from b, this time using only edges not in T. Again the trail must finally come to a halt at b. But then we have obtained a longer cyclic trail from a by following T in a trail T_{ab} to b, then taking the cyclic trail U and returning to b, and finally following the remaining part T_{ba} of T back to a (see Figure 2.6). If we still have not covered the whole graph, we can enlarge the trail again, and so on until we actually have an Eulerian trail. □

The drawing of Eulerian trails is an entertainment familiar to those who work the puzzles in children's magazines. In such puzzles, you are asked to find out how a picture of some kind can be drawn in one continuous line, without repetitions and without lifting the pencil from the paper.

Instead of restricting ourselves to cyclic trails, we often drop the condition that the trail covering all edges shall return to the initial point. When there exists a trail T_{ab}, starting at a and ending at another vertex b, passing once along the edges, then T must depart from the vertex a on some edge and possibly re-enter and re-depart from a a number of times. If this trail does not end at a, then the vertex a must be odd. For an analogous reason b is odd, while other vertices must be even. This yields the following result:

THEOREM 2.2. *A connected graph has a trail T_{ab} covering all edges just once if and only if a and b are the only odd vertices.*

The proof follows from the fact that we can add a new edge ab so that all vertices become even. By the previous theorem, the new graph has an Eulerian trail U, and when the edge ab is dropped from U, the remaining trail is T_{ab}. As an example, we may take the graph in Figure 1.6 which has just two odd vertices f and c and the covering trail *fcdbaec*.

Mathematicians are forever searching for generalizations of the results they have already found. In this spirit, let us try to determine for a general graph the smallest number of trails such that no two of them have a common edge, and all these trails together cover the entire graph. If there is such a family of trails in a graph, we notice that every odd vertex must be the initial point or the endpoint of at least one of them, for otherwise the vertex would have to be even. As we saw in Section 1.6, the number of odd vertices is even, say $2k$. Thus, according to what we just stated, any family of trails T covering the edges must include at least k trails. We show next that the number $2k$ of odd vertices is sufficient for k trails.

THEOREM 2.3. *A connected graph with $2k$ odd vertices contains a family of k distinct trails which, together, traverse all edges of the graph exactly once.*

PROOF. Let the odd vertices in the graph be denoted by

$$a_1, a_2, \ldots, a_k, \quad b_1, b_2, \ldots, b_k,$$

in some order. When we add the k edges $a_1b_1, a_2b_2, \ldots, a_kb_k$ to the graph, all vertices become even and there is an Eulerian trail T. When these edges are dropped out again, T falls into k separate trails covering the edges in the original graph. □

As an example, we may take the graph in Figure 1.1. It has 4 odd vertices, a, b, d, e, and is covered by the two trails *ebfa* and *bcadfed*.

Problem Set 2.3

1. Determine how many trails are necessary to cover the graphs in Figure 2.7.
2. Do the same for all graphs used as illustrations in the preceding pages.
3. Determine covering trails for the complete graphs with 4 and 5 vertices. Try to generalize.

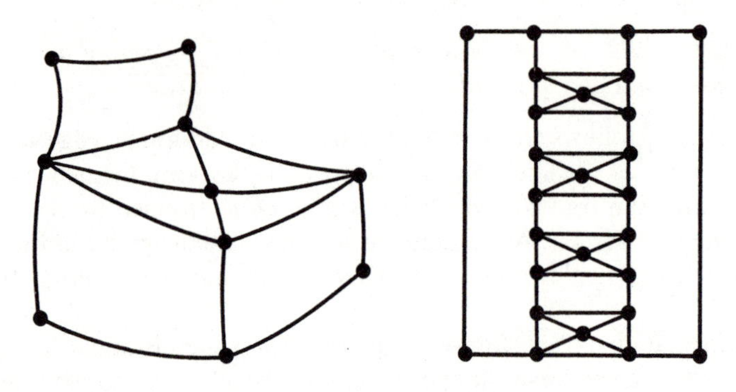

Figure 2.7

2.4 Finding Your Way

An Eulerian graph would be a suitable plan for an exposition, for we can indicate by signs along the pathways how the public should move in order to pass each exhibit once. But suppose that, as usual, the show is so arranged that there are exhibits on both sides of the pathways. Then it is possible, without any restrictions on the graph (except connectedness, of course), to guide the visitor in such a manner that each pathway is traversed twice, once in each direction.

To verify this, we shall describe a general rule for constructing a route that passes along all edges of a graph just once in each direction. We begin our walk along some edge $e_0 = a_0a_1$ from an arbitrary vertex a_0. We mark this edge with a little arrow at a_0 to indicate which direction we have taken. We proceed successively to other vertices; the same vertex may be visited several times. At a_1, and each time later when a vertex is reached, we leave an arrow on the edge to indicate the direction of arrival. In addition, the first time we arrive at a new vertex we mark the entering edge specially so that it can be recognized later.

From each vertex we always exit along unused directions, either along edges which have not previously been traversed or along edges which have been marked as arrival edges; only when there are no other choices may we use the first entering edge as an exit.

We continue this winding walk as far as it is possible. At any vertex there are just as many possibilities for an exit as for an entry. As a consequence, the process can stop only at the initial vertex a_0. It remains to establish that at each vertex all edges have been traversed in both directions.

At a_0 this is simple, for all exit edges must have been used (since otherwise we could have gone further); hence all entering edges have been used, since there are just as many of these. In particular, the edge $e_0 = a_0a_1$ has been covered in both directions. But this means that all exits at a_1 have also been used, since the first entering edge should only be followed as a last resort. The same reasoning applies to the next edge $e_1 = a_1a_2$ and the next vertex a_2, and so on. In this manner we find that at all vertices we have reached all edges are covered in both directions. Since our graph is connected, this means that the whole graph has been traversed.

This method of passing through all edges of a graph may be used for many purposes. It may be used for finding a way out of a maze or a labyrinth, and should you by chance be lost in a cave you may give it a try.

Problem Set 2.4

1. Apply the preceding method to the graphs illustrated in Section 1.1.

2.5 Hamiltonian Cycles

In the year 1859 the famous Irish mathematician Sir William Rowan Hamilton put on the market a peculiar puzzle. Its main part was a *regular dodecahedron* made of wood (Figure 2.8). This is one of

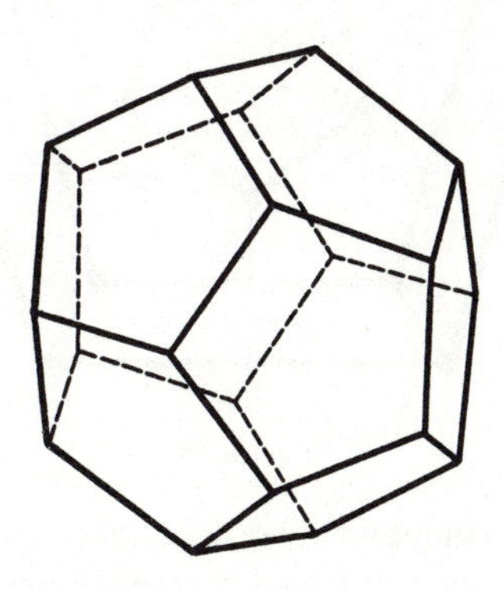

Figure 2.8

the so-called *regular Platonic solids*, and is a polyhedron with regular pentagons for its 12 faces, with three edges of these pentagons meeting at each of the 20 corners.

Each corner of Hamilton's dodecahedron was marked with the name of an important city: Brussels, Canton, Delhi, Frankfurt, and so on. The puzzle consisted in finding a route along the edges of the dodecahedron which passed through each city just once; a few of the first cities to be visited were stipulated in advance to render the task more challenging. To make it easier to remember which passages had already been completed, each corner of the dodecahedron was provided with a nail with a large head, so that a string could be wound around the nails as the journey progressed. The dodecahedron was cumbersome to maneuver, so Hamilton produced a version of his game in which the polyhedron was replaced by a planar graph isomorphic to the graph formed by the edges of the dodecahedron (Figure 2.9).

There is no indication that the Traveller's Dodecahedron had any great public success, but mathematicians have preserved a permanent momento of the puzzle: a *Hamiltonian cycle* in a graph is a cycle that passes through each of the vertices exactly once. It does not, in general, cover all the edges; in fact, it covers only two edges at each vertex. The cycle drawn in Figure 2.9 is a Hamiltonian cycle for the dodecahedron.

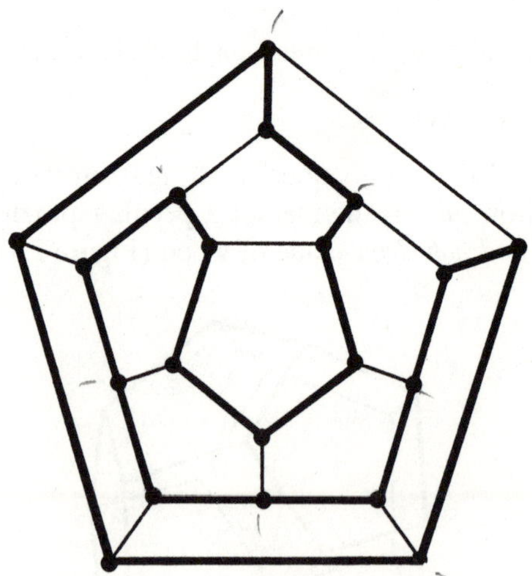

Figure 2.9

There is a certain analogy between Eulerian trails and Hamiltonian cycles. In the former we must pass each edge once; in the latter, each vertex once. In spite of this resemblance the two problems represent entirely different degrees of difficulty. For an Eulerian graph, it is sufficient to examine whether all vertices are even; for Hamiltonian cycles, mathematicians have so far found no such general criterion. This is regrettable, since there are many important questions in graph theory which depend on the existence or non-existence of Hamiltonian cycles. However, it seems that there is *no* efficient general method for determining the existence of such cycles.

The *Travelling Salesman Problem* is a problem in the field of operations research which is reminiscent of Hamiltonian cycles; again we know of no general method of solution. Suppose that a travelling salesman is obliged to visit a number of cities before he returns home. Naturally he is interested in doing this in as short a time as possible, or perhaps he may be concerned about doing it as cheaply as possible. He can, of course, solve the problem by trial and error, finding out the total time, distance or cost for the various possible orders of the cities, but for a large number of stops this becomes unmanageable; for example, if there are 100 cities, then the number of possible routes is about 9×10^{157}, an impossibly large number. Nevertheless, some large-scale examples have been computed, among them the shortest airline distance for a cycle around all the capital cities in the United States.

There are also a number of procedures which, while not giving the best possible solution to the travelling salesman problem, are good enough for most practical purposes. An alternative approach, which gives a *lower bound* for the solution, will be given in Section 3.4.

Problem Set 2.5

1. Do the graphs in Figure 1.1 and Figure 1.2 have Hamiltonian cycles?
2. A salesman lives in the city a_1 and has to visit the cities a_2, a_3, a_4. The distances (in miles) between these cities are

$$a_1a_2 = 120, \qquad a_1a_3 = 140, \qquad a_1a_4 = 180,$$

$$a_2a_3 = 70, \qquad a_2a_4 = 100, \qquad a_3a_4 = 110.$$

Find the shortest round trip from a_1 through the other three cities.

2.6 Puzzles and Graphs

Previously we discussed how to find the way from one place to another in a graph. This problem may be considered to be a sort of game, and in spite of the fact that it appears to be quite a simple-minded pastime, it actually represents the main content of many puzzles and games.

Let us use the very ancient *Ferryman's Puzzle* to illustrate what we have in mind. A ferryman (f) has been charged with bringing across a river a dog (d), a sheep (s), and a bag of cabbage (c). His little rowboat can carry only one of the items at a time; furthermore, he cannot leave the dog alone with the sheep, nor the sheep with the cabbage. How shall he proceed?

We analyze the various possible alternatives. The only permissible first move is to bring the sheep over; this changes the group at the starting point from f, d, s, c to d, c. He then comes back alone, making it f, d, c. Next he can take either d or c across, leaving c or d. In either case he must take s back, giving f, s, d or f, s, c at the starting point, as the case may be. On his next trip he takes d (or c) across, leaving only s. Finally he comes back alone and transports s across.

Thus in this extremely simple case we have only the permissible moves which are indicated in the graph in Figure 2.10; the items carried at each stage are indicated on the edges of the graph. Thus the solution can be reached in two ways, each by a path from the initial position f, d, s, c to the final position "none."

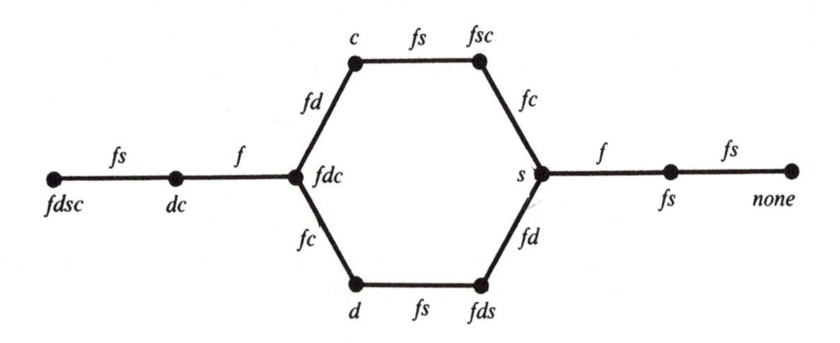

Figure 2.10

A problem of a similar type is the *Puzzle of the Three Jealous Husbands*: three married couples on a journey come to a river where they find a little skiff which cannot take more than two persons at a time. The crossing is complicated by the fact that the husbands are all very jealous and will not permit their wives to be left without them in a company where there are other men present. We leave it to you to draw the graph of the permissible moves and show how the transfer can be effected (see Problem 1, below).

As we have seen in the preceding examples, we can conceive of a graph as a game. The vertices are the various positions in the game and the edges represent the moves which are permitted according to the rules. A usual problem is to decide whether or not we can move from one given position to another stepwise, along edges of the graph. In the language of graph theory, this becomes the question: are the two positions in the same connected component of the graph?

As a further simple example, let us consider for a moment a game consisting in moving the knight of a chess game around the board according to the usual rule—that is, two squares horizontally or vertically and one square in a perpendicular direction. Since there are 64 squares on the board, the corresponding graph has 64 vertices. It is not difficult to see that the knight can reach any square from any original position, so the game graph is connected.

In some of the earliest manuscripts on chess, one runs across the following question: is it possible to move the knight from some arbitrary starting position around the whole board and return it to the starting point so that each square has been occupied just once? This is the same as finding a Hamiltonian cycle for the graph. There are in fact many solutions; one of them is given in Figure 2.11.

63	22	15	40	1	42	59	18
14	39	64	21	60	17	2	43
37	62	23	16	41	4	19	58
24	13	38	61	20	57	44	3
11	36	25	52	29	46	5	56
26	51	12	33	8	55	30	45
35	10	49	28	53	32	47	6
50	27	34	9	48	7	54	31

Figure 2.11

There are a large number of different moves the knight may make from one square to another. We may ask whether it is possible to find a cyclic route which includes them all just once. This corresponds to the construction of an Eulerian trail in the graph and so, according to our general result, we must examine whether the degrees of the vertices are all even. In Figure 2.12 we have indicated for each square how many possible knight's moves there are—that is, the degrees of the vertices of the graph. There are 8 squares with the odd degree 3, so the graph has no Eulerian trail.

2	3	4	3	2
3	4	6	4	3
4	6	8	6	4
3	4	6	4	3
2	3	4	3	2

Figure 2.12

Problem Set 2.6

1. Solve the above puzzle of the jealous husbands.
2. Prove that the puzzle of the jealous husbands cannot be solved for 4 couples, but that it can be solved if the ferryboat holds 3 persons.
3. How many moves of the knight are there on the chess board?
4. Solve the corresponding problems for the moves of the king.
5. Verify that the numbers in each row and the numbers in each column in Figure 2.11 give the same sum, 260.

CHAPTER THREE

Trees

3.1 Trees and Forests

A *tree* is a connected graph that has no cycles. This means in particular that there are no multiple edges. It also implies that in a tree there is a unique path connecting any pair of vertices. Graphs without cycles have connected components which are trees; this makes it natural to extend the botanical terminology, and call such graphs *forests*.

To construct a tree, we select some particular vertex a_0. From a_0, we draw edges to neighboring vertices $a_1, a_2, \ldots$; from these we draw edges to their neighbors $a_{11}, a_{12}, \ldots, a_{21}, a_{22}, \ldots$ and so on, as indicated in Figure 3.1. The particular vertex a_0 which we have chosen in Figure 3.1 is called the *root* of the tree; any vertex could have been used as the root.

Since there are no cycles in the tree, the various paths (or branches) from a_0 must remain apart once they have become separated, just like the branches of an ordinary tree. Each branch in the graph must have a last *terminal edge* to a *terminal vertex* from which there are no further edges.

According to this observation, we can also construct the tree by successively hanging on edges at the vertices. This makes it possible to tell how many edges there are in a tree. The simplest tree is a single

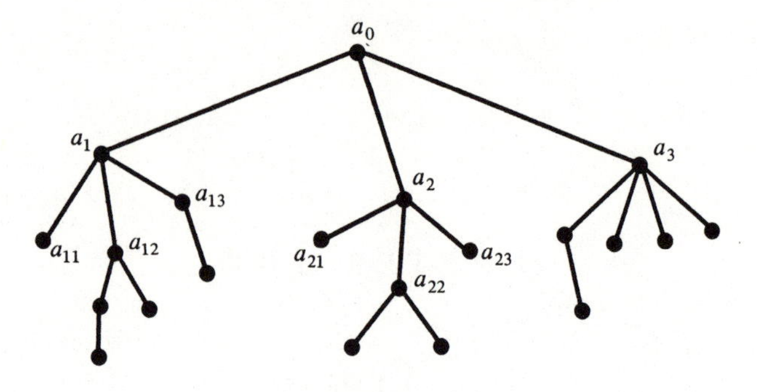

Figure 3.1

vertex; it has no edges. Each time an edge is added at the end of a branch one also adds a vertex, so that we can conclude:

THEOREM 3.1. *A tree with n vertices has* $n - 1$ *edges.*

Instead of taking a tree, we could have considered a forest with k connected components, all trees (Figure 3.2). In each component tree there is one edge fewer than the number of vertices, so that we may state:

THEOREM 3.2. *A forest with k components and n vertices has* $n - k$ *edges.*

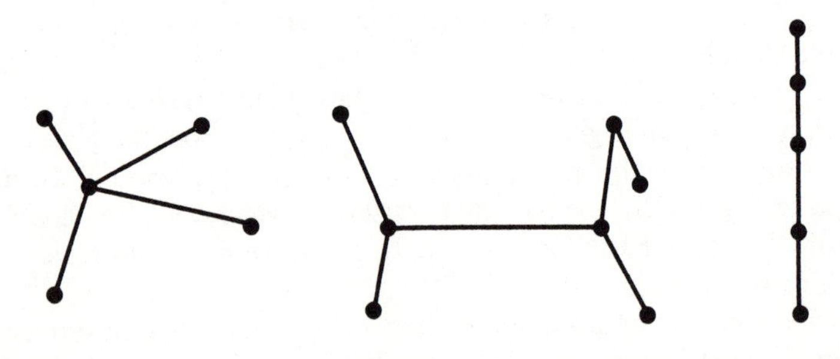

Figure 3.2

There are many applications of trees in fields as wide-ranging as chemistry, linguistics and computing. At this stage, let us mention only that any sorting process can be pictured in the form of a tree. For instance, we may consider Figure 3.1 to be the result of a mail sorting. An original bunch of letters is placed at a_0. The domestic mail may be sorted to a_1, the mail for Europe to a_2, the mail to the Far East to a_3, and so on. The domestic mail at a_1 is sorted next to Eastern, Western, and Central mail; the European mail at a_2 may be sorted according to countries, and so on.

The Dewey decimal classification of books (as used in libraries) may be represented by a graph in the same way, only here the tree takes on a quite regular shape. When we sort according to the basic classification (pure sciences, applied sciences,...), there are 10 alternatives, $a_0, a_1, a_2, \ldots, a_9$; for each of these there will again be 10 alternatives, such as $a_{00}, a_{01}, \ldots, a_{09}$, and so on (see Figure 3.3). The whole procedure can be considered to be a sorting of the numbers from 0 to 999 according to the first, second,... digit. Indeed, it is possible to conceive of any tree as a kind of very general number system.

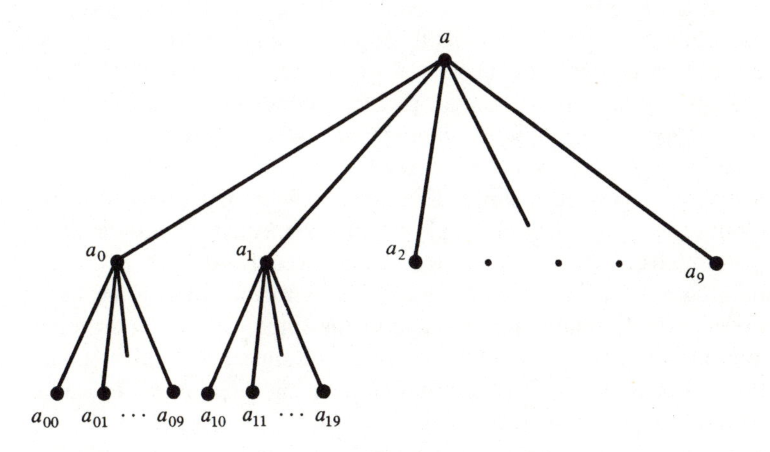

Figure 3.3

3.2 Cycles and Trees

Let us formulate our next problem in agricultural terms. In Figure 3.4 we have drawn a map of farm fields. We shall think of this map as representing a number of rice fields on an island; the fields are surrounded by earthen dams and these, in turn, are surrounded by the waters of a lake.

As is usual in rice cultivation, we want to set the fields under water by opening up some of the walls. In order to immerse every field, we

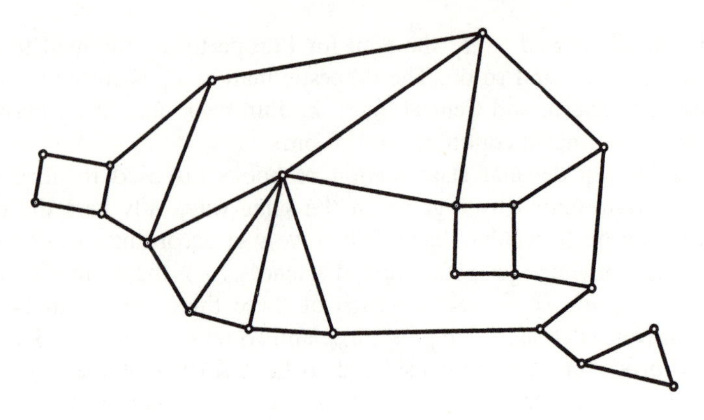

Figure 3.4

must evidently break at least one wall in each cycle of the map—that is, the remaining unbroken walls must give us a graph without cycles. The question is then: how many walls is it necessary to pierce?

This brings us to a general problem concerning graphs: *in a connected graph, what is the smallest number of edges that must be removed in order that no cycles remain*?

Suppose that we first eliminate an edge $e = ab$ belonging to some cycle in the graph. Then the graph remains connected, because instead of passing from a to b on e, we can proceed from a to b on the remaining part of the cycle. If there are further cycles after e has been removed, we eliminate another edge in the same way. By continuing in this manner, we must finally come to a connected graph without cycles—that is, a tree T. (Note that eliminating different edges leads, in general, to different trees.)

When we have arrived at this stage, it is simple to determine the number of edges which have been removed. The tree T has the same number n of vertices as the original graph G. According to Theorem 3.1, there are $n - 1$ edges in T. Therefore, if G originally had m edges we have removed exactly

$$\gamma = m - n + 1$$

edges. This number is called the *cycle rank* of the graph G. It is the difference between the number of edges and the number of vertices of G, increased by 1.

We have established that, in order to reduce a graph G to a tree, we must always remove at least γ edges. In order to reduce G to a forest consisting of several trees, one must always remove more than γ

edges, since (according to Theorem 3.2) a forest with n vertices has fewer edges than a tree with n vertices.

Let us illustrate the reduction on the graph in Figure 1.1. Here the edge *ed* belongs to the cycle *efd* and we remove it first. The edge *ad* belongs to the cycle *dfa* and is removed. Finally, *ac* and *be* are eliminated. This leaves us with the tree in Figure 3.5. Note that we have removed

$$\gamma = 9 - 6 + 1 = 4$$

edges.

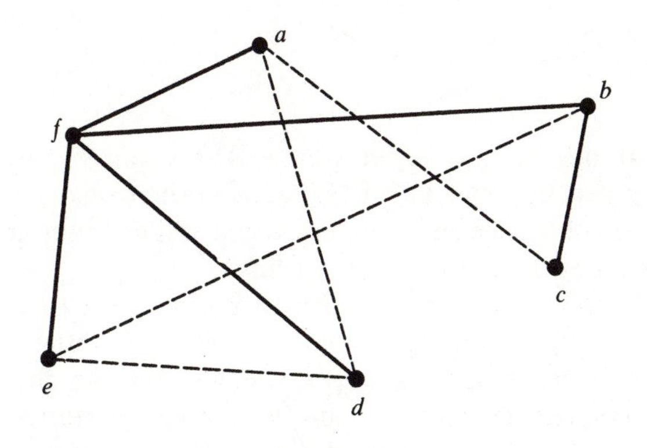

Figure 3.5

Problem Set 3.2

1. Find the cycle ranks of the graphs in Figure 1.2 and Figure 3.4.
2. What is the cycle rank of the complete graph K_n?

3.3 The Connector Problem

We now turn to a communication problem that has some very practical uses and pose it first in the form of a road construction question. We have a certain number of cities $a, b, c, \ldots$ and we want to construct a road or railroad net connecting all of them. For each pair of cities a, b we know the cost $c(ab)$ of constructing a connecting line between them. The problem is to build the whole network as cheaply as possible. Instead of using railroads, one can illustrate the situation by means of electrical wire connections, or water mains, or gas and oil pipes.

In the special case where there are only three cities a, b and c it is sufficient to build one of the connecting lines abc, acb, bac (see Figure 3.6). If bc is the most expensive stretch then it should be left out and the connecting links bac should be built.

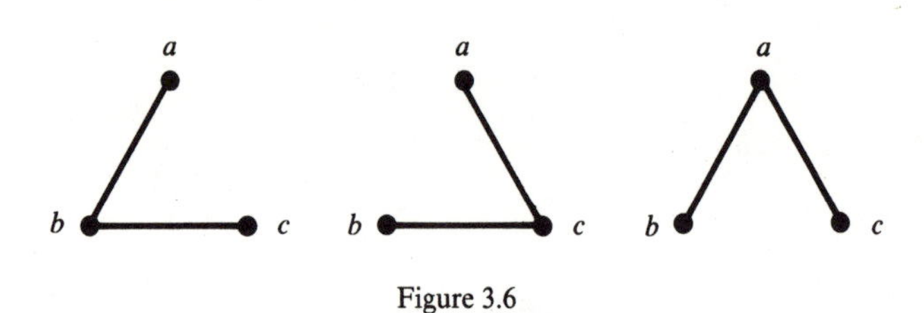

Figure 3.6

We turn next to the general case. The graph of the cheapest connecting network must be a tree, because otherwise one could leave out any edge of a cycle and the cities would still be connected. Thus if one has n cities there must be $n - 1$ links.

We shall show that a minimum cost network can be constructed according to the following simple *economy rule* (often called the *greedy algorithm*): in the first step we connect the two cities with the cheapest connecting link e_1; in each step thereafter we take the cheapest possible edge e_i producing a tree together with the edges already selected; if there should be several edges with the same cost, it does not matter which one is used. Any tree T constructed in this manner may be called an *economy tree*. Its cost $c(T)$ is the sum of the costs for the various edges;

$$c(T) = c(e_1) + c(e_2) + \cdots + c(e_{n-1}).$$

There remains the essential point: to demonstrate that no other connecting tree S can have a lower cost $c(S)$ than an economy tree. Take S to be a connecting tree with the smallest cost, and T any economy tree. Suppose the edges $e_1, e_2, \cdots$ of the economy tree T are numbered in the order in which they were added in the construction of T. If the minimal cost tree S is not identical with T, then T has at least one edge not in S; let $e_i = ab$ be the first edge of T not in S, and let P_{ab} be the path in S connecting the vertices a and b (see Figure 3.7). If the edge e_i is added to S, the graph $S + e_i$ will have a cycle $C = e_i + P_{ab}$ and since T has no cycle, C must contain at least

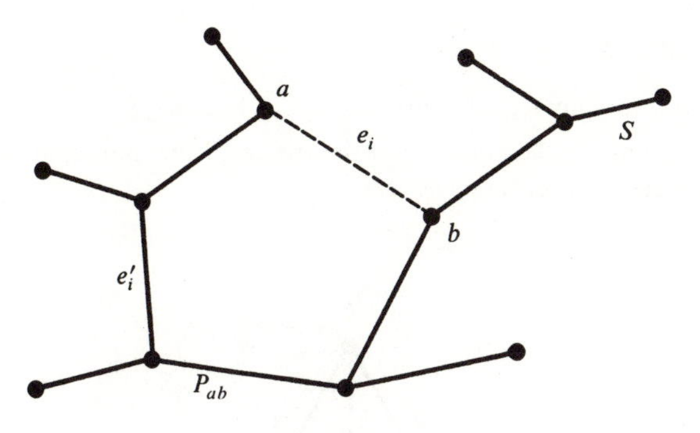

Figure 3.7

one edge, say e_i', not in T. We remove this edge and obtain the tree

$$S' = S + e_i - e_i'$$

with the same vertices as S and whose cost is

$$c(S') = c(S) + c(e_i) - c(e_i').$$

Since S has the smallest possible cost,

$$c(e_i) \geqslant c(e_i').$$

But e_i was the edge with the smallest cost such that when added to $e_1, e_2, \ldots, e_{i-1}$ no cycle was produced. Since e_i' when added to these edges does not give any cycle we conclude that

$$c(e_i) = c(e_i'),$$

and so S' has minimum cost

$$c(S) = c(S').$$

In this manner we have found another tree S' with minimum cost and one more edge, namely e_i, in common with the economy tree T. But then we can repeat this operation until we finally obtain a connecting tree with minimum cost which coincides with T. Thus T and all other economy trees have minimum cost.

This method for constructing economy trees is very efficient in practice, even for road networks with a large number of cities. In particular, it is very suitable for computer implementation.

Problem Set 3.3

1. Find two economy trees for the network in Figure 3.8.
2. Draw 6 points in the plane. Find the tree with the minimum total length whose edges connect these vertices.

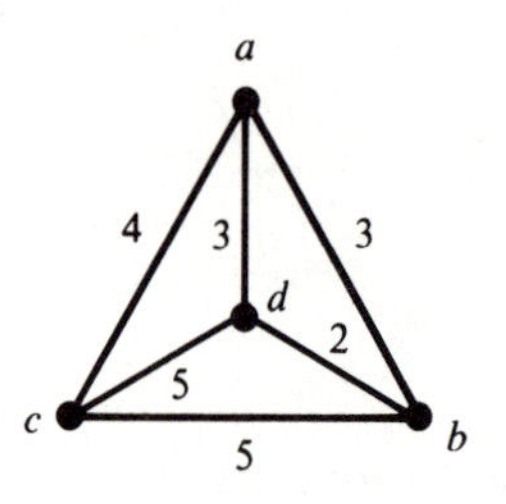

Figure 3.8

3.4 The Travelling Salesman Problem Revisited

In Section 2.5 we mentioned the *travelling salesman problem*, in which a travelling salesman wishes to visit a number of cities and then return home, incurring the smallest possible total cost. Since the number of cyclic routes joining the cities is usually far too large for a trial-and-error approach, a systematic method is needed. Surprisingly, the economy rule of the previous section can be used to give a lower bound for the solution of the travelling salesman problem. Let us see why this is.

Suppose that the cycle on the left of Figure 3.9 gives a solution of the travelling salesman problem for the five cities a, b, c, d and e. If

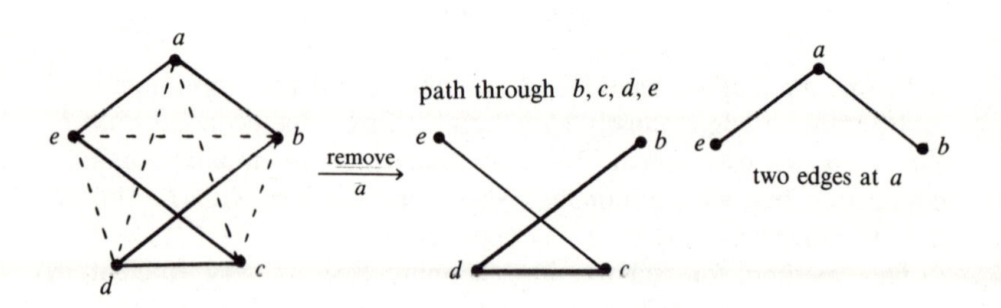

Figure 3.9

we remove the vertex a, we get a path through the vertices b, c, d and e. Such a path is a tree through these vertices, and the total cost incurred by the salesman is obtained by adding the costs of the edges in this tree to the costs of the two edges at a. It follows that, if we add the cost of an *economy* tree through b, c, d and e to the two *smallest* costs at a, we cannot exceed the solution of the travelling salesman problem. In other words, *we can obtain a lower bound for the solution of the travelling salesman problem by adding the lengths of the economy tree through* b, c, d *and* e *to the two smallest costs at* a.

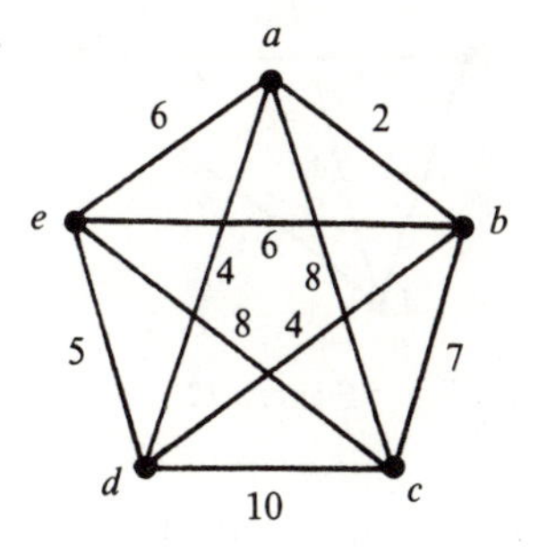

Figure 3.10

As an example of the use of this method, consider the five cities in Figure 3.10. If we remove the vertex a, then we get the following graph (Figure 3.11) with vertices b, c, d and e. The economy tree through these vertices is the tree with edges bc, bd and de, with total cost 16. The two smallest costs at a are 2 and 4 (for the edges ab and ad). The required lower bound for the solution of the travelling salesman problem is thus $16 + 2 + 4 = 22$.

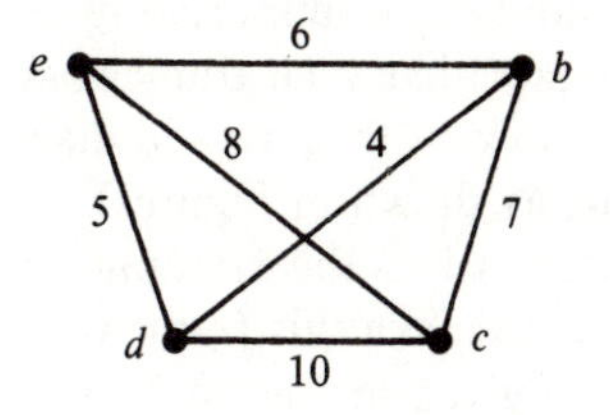

Figure 3.11

We can obtain a better lower bound by removing the vertex c. In this case, the remaining graph has vertices a, b, d and e (Figure 3.12) and there are two economy trees, each with total cost 11—namely, the trees with edges ab, ad or bd, and de. The two smallest costs at c are 7 and 8 (for the edges cb, and ca or ce). The required lower bound for the solution of the travelling salesman problem is thus $11 + 7 + 8 = 26$, a definite improvement on the bound obtained by removing vertex a.

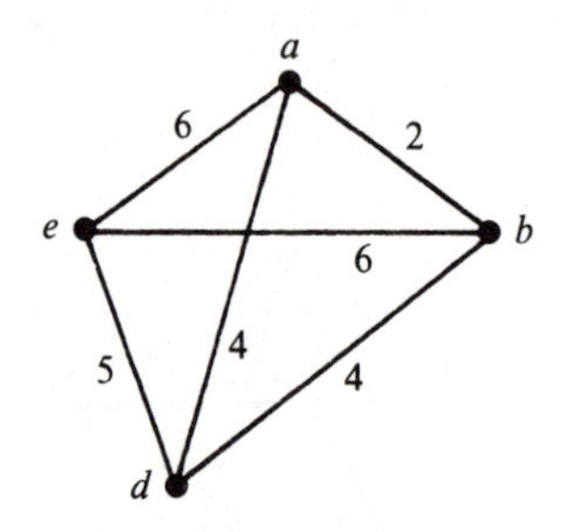

Figure 3.12

It follows from these results that the solution to the travelling salesman problem is at least 26. In fact, as a little experimentation will convince you, removing the vertex c actually gives the correct answer in this case—the cyclic route of minimum cost through the vertices a, b, c, d and e is $abceda$, with the total cost $2 + 7 + 8 + 5 + 4 = 26$.

Problem Set 3.4

1. In the above travelling salesman problem, what are the lower bounds obtained by removing
   (i) the vertex b? (ii) the vertex d? (iii) the vertex e?

3.5 Bracing Frameworks

We now use the properties of trees to solve a problem in structural engineering. Many buildings are supported by rectangular steel frameworks, and it is important that such frameworks should remain rigid under heavy loads. To prevent this, we add diagonal braces to prevent distortion in the plane, as shown in Figure 3.13. (The diagonal braces can go in either direction.) Is 3 the minimum number of such braces that wc must add so as to make this framework rigid?

Further examples are given in Figure 3.14. Framework (i) is rigid, but is over-braced since several of the diagonal braces can be removed

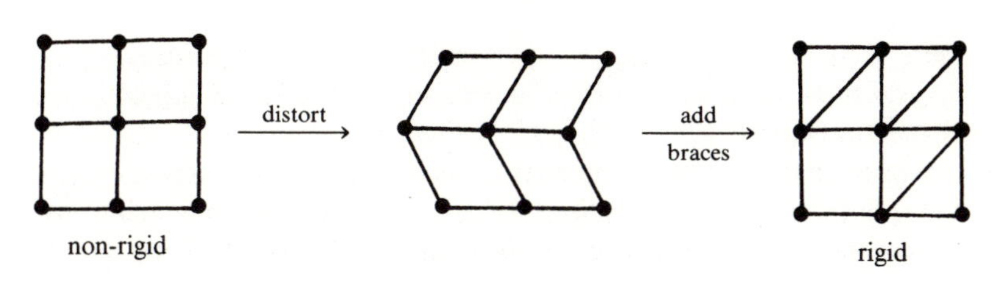

Figure 3.13

without affecting the rigidity of the framework. Framework (ii) is not rigid, since it can be distorted as shown. But how about framework (iii)? Is it rigid? If so, can any diagonal braces be removed without affecting the rigidity?

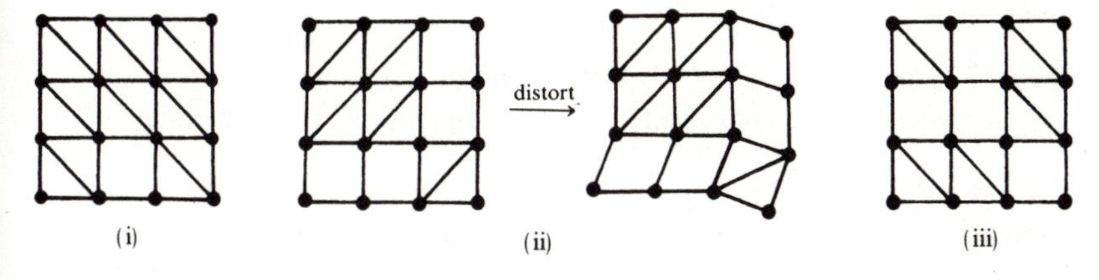

Figure 3.14

We can determine whether a given braced framework is rigid by drawing a graph whose vertices correspond to the rows and the columns of the framework, and whose edges join a row-vertex and a column-vertex whenever there is a diagonal brace in the corresponding row and column. For example, the graphs corresponding to the frameworks of Figure 3.14 are as shown in Figure 3.15; in each case

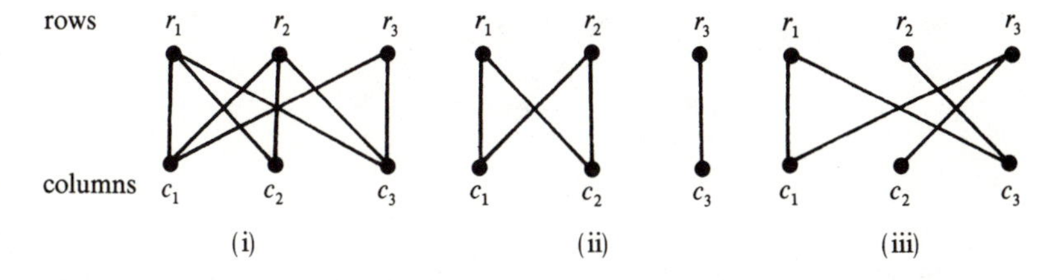

Figure 3.15

there are three row-vertices r_1, r_2, r_3 and three column-vertices c_1, c_2, c_3, and there is an edge from r_1 and c_1 since there is a diagonal brace in the r_1c_1-square.

Notice that the rigid framework (i) gives rise to a connected graph, whereas the non-rigid framework (ii) gives rise to a graph which is not connected. This is because each diagonal brace forces the corresponding row and column to be at right angles. Thus, in the graph of framework (i), the edges from r_1 and r_2 to c_1, c_2 and c_3 show that rows 1 and 2 are at right angles to all three columns, and similarly row 3 is at right angles to columns 1 and 3; it follows that every row is at right angles to every column, and thus the framework cannot be distorted. However, in the graph of framework (ii), the vertices r_3 and c_3 are not connected to the other vertices; thus row 3 and column 3 need not remain at right angles to the other rows and columns, and the framework can be distorted. These examples are instances of the following general rule, which shows that framework (iii) is also rigid since its graph is connected:

if the graph is connected, then the framework is rigid;
if the graph is not connected, then the framework can be distorted.

Notice also that the graph of framework (i) contains several cycles —for example, $r_1c_2r_2c_1r_1$. If we remove any edge from this cycle, the graph remains connected, and so the framework remains rigid. We can continue removing edges from the cycles in the graph until there are no cycles left—for example, if we successively remove the edges r_1c_1, r_2c_2 and r_3c_1, the graph remains connected, and so the framework remains rigid. The remaining graph is a tree through all the vertices, called a *spanning tree* (see Figure 3.16). The removal of any further edge would disconnect the tree, and the framework would cease to be rigid.

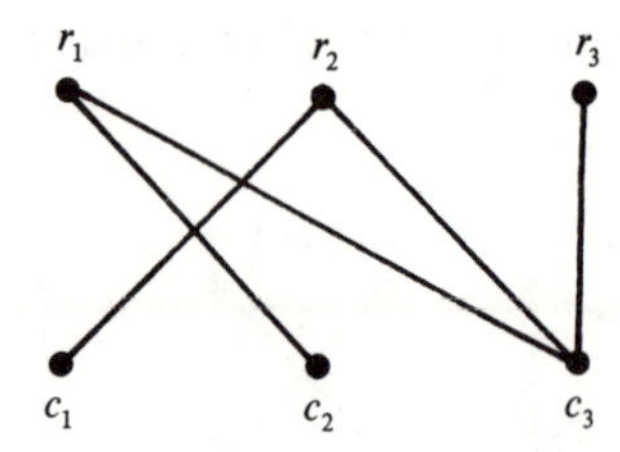

Figure 3.16

Thus we have the following rule, which shows that the frameworks in Figure 3.14(iii) and on the right of Figure 3.13 are *minimum bracings*—that is, no further diagonal braces can be removed without affecting the rigidity of the framework:

if the graph is a spanning tree, then the bracing is a minimum bracing.

Problem Set 3.5

1. Which of the braced frameworks in Figure 3.17 is rigid? Which of them has a minimum bracing?
2. If the framework has m rows and n columns, how many diagonal braces are needed for a minimum bracing?

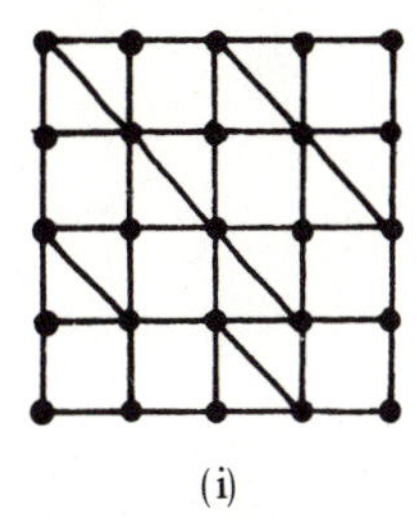

(i)

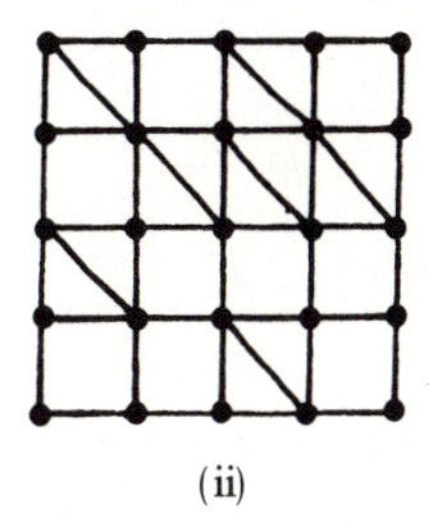

(ii)

Figure 3.17

3.6 Streets and Squares

Changing the names of streets and public squares has been a favorite pastime for city councils all over the world. Suppose now that the city fathers want to be very systematic about their street names. Each street shall be a block long and each street shall carry the same name as one of its adjoining street intersections; so, for instance, Washington Avenue or Street shall have one of its ends at Washington Square.

We naturally want to put the question in a general graph form. A connected graph is given. *When is it possible to let each edge correspond in a unique manner to one of its endpoints*?

We point out to begin with that this is always the case when the graph is a tree. After we have selected an arbitrary root a_0 in the tree as in Figure 3.1, we let the edge a_0a_1 correspond to the vertex a_1, similarly a_0a_2 to a_2, and a_0a_3 to a_3. In the next step a_1a_{11} corresponds to a_{11}, and so on; in general, an edge corresponds to its endpoint farther away from a_0.

Suppose next that the graph has a cycle C (Figure 3.18). Each edge in C corresponds to one of its endpoints—that is, to a vertex on C. If, for instance, the edge a_1a_2 corresponds to a_2, then the edge a_2a_3 must correspond to a_3, and so on. No edge not in C can correspond to a vertex in C.

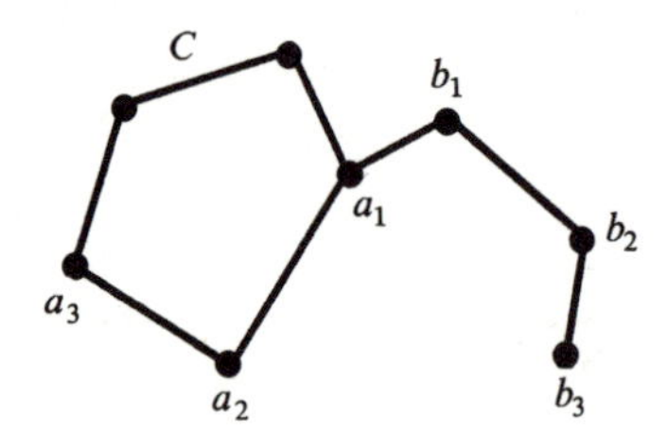

Figure 3.18

As a consequence, any edge a_1b_1 touching C at a_1 must correspond to b_1, and any edge b_1b_2 to b_2, and so on. But such a path $a_1b_1b_2b_3\ldots$ cannot return to C because these vertices have already been matched with edges of the cycle; nor can the path run back into itself for the same reason.

We see therefore that the part of the graph which we can reach from a_1 beginning in edges a_1b_1 touching C must be a tree T_1 with the root a_1, and similarly for every other vertex a_i on C. But we have just noted that in a tree T_1 we can let each edge correspond to that endpoint which is farther away from the root a_1. Since the edges of C correspond to the vertices a_i we obtain as a result of this analysis:

in a connected graph we can let each edge correspond uniquely to one of its endpoints if and only if the graph is a tree or it consists of a single cycle C with trees growing at its vertices (see Figure 3.19).

According to Theorem 3.1, a tree has one more vertex than it has edges; a cycle, or a cycle with trees growing from its vertices has the same number of edges and vertices. Thus the possibility of matching edges with vertices is to be expected in these cases.

A tree as in Figure 3.1, or a graph as in Figure 3.19, can represent a street map only for very small towns where there are no real city

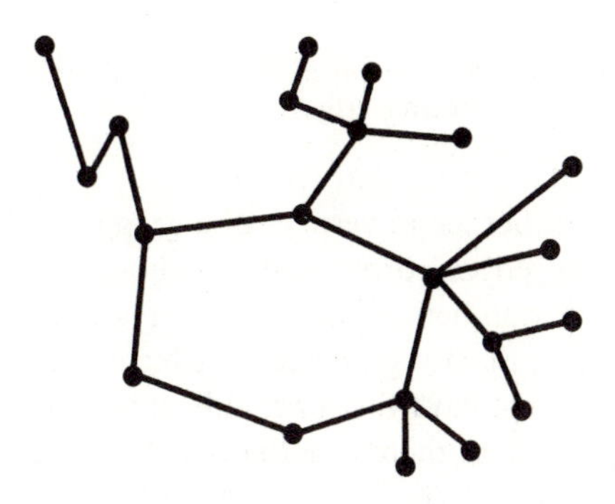

Figure 3.19

blocks, or where there is only one central block to which the roads lead in from the country.

After reflecting a little upon the situation, the city council members proudly become aware of the fact that their town is much too large for the use of this method. Instead they agree to substitute the following principle: the streets and the squares shall be so named that at each square there is a street with the same name—that is, for instance, at Washington Square there must always be a Washington Street running into it.

In terms of graph theory, this means matching each vertex to a unique adjoining edge. There are cases in which this cannot be done; a tree, for example, has one more vertex than it has edges, as we saw in Theorem 3.1. But here it can almost be done. Let us look at the illustration of a tree in Figure 3.1. One can associate with each vertex that edge which leads from it to the root a_0. This matches each vertex to an edge as desired, except for the root itself.

The trees are exceptions in this respect for we have the general result:

in a connected graph which is not a tree, we can always match the vertices to adjoining edges.

PROOF. A graph connecting a set of n vertices has at least $n - 1$ edges; if it is not a tree, it has more than $n - 1$ edges and can be reduced to a tree by having some edges removed (see Section 3.2). Let $e_0 = a_0b_0$ be one of the edges removed when the graph is reduced to a tree T, and select a_0 as its root. In T, every vertex except a_0 can be made to correspond to an adjoining edge; the extra edge e_0 of our graph can be assigned to a_0, and now every vertex of the graph is matched with an adjoining edge. □

It is of interest to note that, according to the preceding discussion, we can always either let all vertices of a graph correspond to adjoining edges or let the edges correspond to adjoining vertices. We can do both when the number of vertices of a graph is the same as the number of edges. Such a graph cannot be a tree, and so we conclude that the graph must have the form indicated in Figure 3.19, with only a single cycle C. Here there is actually a correspondence which works both ways, edges to vertices, and vertices to edges. One lets the edges on C correspond to vertices on C, while any other vertex not on C corresponds to one of its edges nearest to C.

Problem Set 3.6

1. Give a correspondence of vertices to adjoining edges in the graphs in Figures 1.1 and 1.2.

CHAPTER FOUR

Matchings

4.1 The Jobs and the Applicants

A firm has a number of vacant jobs of various types, and a group of applicants to fill them. Each applicant is qualified for certain of the jobs and so the question comes up: is it possible to assign all the applicants to positions for which they are suited? (This is often called the *Assignment Problem*.)

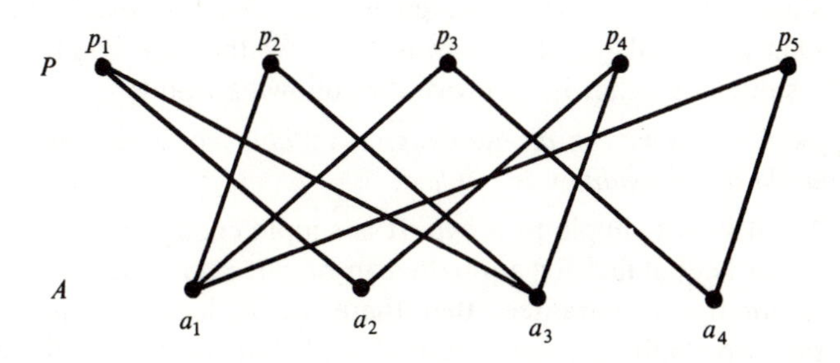

Figure 4.1

We can again illustrate the situation by means of a graph, this time of a somewhat special type. As we have just explained, there is a group of available applicants which we shall denote by A, and a set of positions P. We construct our graph by drawing edges (a, p) connecting each applicant a in A with the positions p in P for which they are qualified. Thus there are no edges connecting two vertices corresponding to the applicants, nor are there any edges connecting two jobs. Therefore, the graph has the form which we have indicated in Figure 4.1. A graph of this kind, where the set of vertices is decomposed into two separate parts A and P such that there are edges only between A and P is called a *bipartite graph* (Figure 4.1).

It is clear that we cannot always expect to have a suitable job for each applicant. For one thing, there must be at least as many jobs as there are applicants. But this is not enough. Imagine, for example, that the applicants consist of two carpenters and an applicant who can do both woodwork and plumbing, and that there are four job openings for these three applicants, one in carpentry and three in plumbing. Then, clearly, one of the carpenters will remain jobless, although there are more jobs than applicants, and although our three applicants, collectively, qualify for both available trades.

Suppose there are altogether n applicants applying for positions. Then in order to be able to solve our assignment problem, the following condition will have to be fulfilled:

for each group of k applicants (for all $k = 1, 2, \ldots, n$), there must be at least k jobs for which, collectively, they are qualified.

For example, if one applicant were a carpenter and the other a carpenter who can also do plumbing, and if two plumbing jobs were available, then this condition would be violated for $k = 1$, although it holds for $k = 2$, and the applicants could not be placed.

We shall call the italicized statement the *diversity condition*, for short. Our principal purpose now is to show that the condition is actually sufficient—that is, to prove the following result:

THEOREM 4.1. *One can always assign a suitable job to each applicant when the diversity condition is fulfilled.*

The result is not simple to prove. It is simple enough for $n = 1$: if an applicant is qualified for a job, he can be put into it. If $n = 2$, the diversity condition guarantees that there are at least two jobs for which the two applicants are qualified, and that, moreover, the carpenter-plumber catastrophe does not occur—that is, that each applicant is qualified for at least one job. Once again the applicants can be placed.

These remarks show that the theorem is true for $n = 1$ and for $n = 2$. It is reasonable to try to prove the theorem in general by the principle of mathematical induction: we shall assume that the statement is true when there are $n - 1$ applicants or fewer, and we shall deduce that it is true when there are n applicants.

If the diversity condition holds, it may hold with room to spare, or just barely. It may, that is, happen that every possible group of k applicants ($k = 1, 2, \ldots, n - 1$) qualifies for more than k jobs (room to spare), or it may happen that for some k_0 ($k_0 = 1, \ldots, n - 1$) there is a group of k_0 applicants who qualify for exactly k_0 jobs (just barely). We shall prove that in either case we can always assign a suitable job to each of the n applicants.

If the diversity condition holds with room to spare, pick any one of the n applicants and place him in one of the jobs for which he is suited. Among the remaining $n - 1$ applicants, no group of k (for any k) can be qualified for fewer than k jobs, because among the originally available jobs there were at least $k + 1$ jobs that these applicants were suited for, and the assignment already made took at most one job opportunity away from them. By the induction assumption, the $n - 1$ applicants can be put into suitable jobs.

If the diversity condition holds just barely, consider some set A_0 of k_0 applicants ($k_0 < n$) who are qualified for just k_0 jobs. Because of the diversity condition, no k of these applicants (for $k = 1, 2, \ldots, k_0$) can be qualified for fewer than k jobs; it follows from the induction assumption that the k_0 applicants in A_0 can be placed. It remains to consider the $n - k_0$ applicants who are left. We must prove that the diversity condition is still satisfied for them and for the unfilled jobs —that is, that for any group B of k of these applicants ($k = 1, 2, \ldots, n - k_0$) there are still at least k unfilled jobs. To see this, suppose that the applicants in group B qualify for only k' jobs, where k' is less than k. Then the set $A_0 + B$ consisting of the $k_0 + k$ applicants in A_0 and in B would originally have been qualified for only $k_0 + k'$ jobs; this is contrary to the assumed validity of the diversity condition. We conclude that the diversity condition holds for the remaining $n - k_0$ applicants, who can therefore be placed in suitable jobs by the induction assumption; this completes the proof of the theorem. □

Figure 4.2 shows the case $n = 6$. The first three vertices on the lower line represent a group A_0 of $k_0 = 3$ applicants qualified, respectively, for plumbing (p), carpentry and plumbing (cp), and carpentry (c), and the edges lead to positions (vertices on the upper line) filled by these applicants. The remaining three vertices represent $n - k_0 = 3$ applicants qualified, respectively, for bricklaying (b), plumbing (p), and toolmaking (t). We observe that the diversity condition is violated for a group B of $k = 2$ of these applicants, the bricklayer and the plumber; there is no plumbing job among the unfilled positions, and only one bricklaying job. Clearly, the diversity condition is

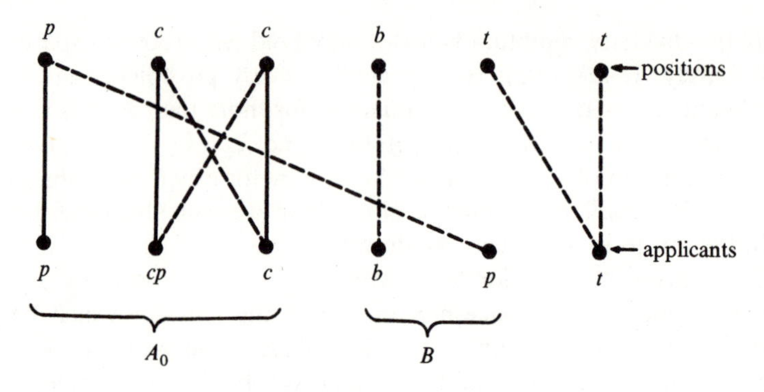

Figure 4.2

therefore also violated for the set $A_0 + B$ of five applicants, because there were to begin with only four positions for which they qualified.

4.2 Other Formulations

When we introduced a bipartite graph, as for instance in Figure 4.1, we considered the one set A of vertices to be the applicants and the other set P as the available positions. Assume now that one can assign a suitable job to each applicant. This means in terms of the graph that one can find one edge at each vertex in A such that each of these edges goes to a different vertex in P. We therefore say that we have a *graph matching* of the vertices in A into the vertices in P (Figure 4.3).

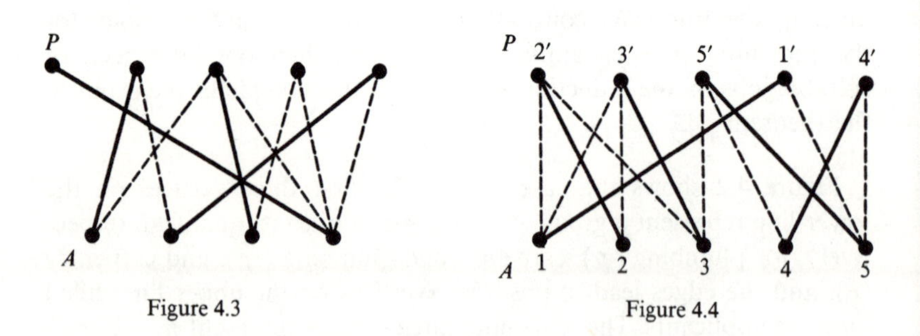

Figure 4.3

Figure 4.4

We saw that such a matching of A into P is possible if and only if the diversity condition is fulfilled—that is, any k vertices in A are connected by edges to at least k vertices in P. If P and A have the same number of vertices, then such a matching gives a one-to-one correspondence between the vertices in A and P, such that corresponding vertices in the matching are connected by a graph edge (Figure 4.4).

This problem of finding graph matchings has been formulated in many ways. Sometimes it is called the *Marriage Problem*: suppose that we have a group of boys and a group of girls, each with the same number of individuals. Some of them know each other already, and the question is: when is it possible to pair the two groups in such a manner that the boy and the girl in each pair are already acquainted with each other?

One can also put the problem in reverse and state it as follows: in a small village there are the same number of boys and girls of marriageable age. By taboos and custom the boys are not permitted to marry close relatives: sisters, half-sisters or cousins. When is it possible for all the boys to take a bride within the village? As before we can depict the possibilities for marriage for each individual by means of a bipartite graph, only in this case they should be connected by an edge when they are not related.

Let us give another twist to our problem. In your class in school there will be a certain number of committees; we may name them

$$C\colon C_1, C_2, \ldots, C_n.$$

Naturally each of these n committees must have a secretary. In order to avoid too dominating an influence by a small group, it has been stipulated that no member shall be permitted to be secretary of more than one committee. Again we have the question: under what conditions is this possible? It is not always possible; when there are too many committees in a relatively small class, we may be in trouble with this rule.

To solve the problem, we turn to the bipartite graphs as before. In this case, one of the vertex sets C in the graph consists of the n committees, and the other vertex set P consists of the pupils in the class. We draw an edge from a committee C_i to a pupil p only if p is a member of C_i. In this case the diversity condition runs as follows: any group of k committees ($k = 1, 2, \ldots, n$) must include at least k distinct pupils. According to our theorem, this is the condition that it be possible to select separate secretaries.

After we have put our problem in this form, we have in reality deduced a theorem which was published by the English mathematician Philip Hall in 1935: a number n of sets C are given; each has members p from a set P. We wish to assign to each set C_i one of its members p_i such that different sets have been assigned a different element in P. This is possible only if the following condition is fulfilled: any k sets C_i $(k = 1, 2, \ldots, n)$ must include at least k different elements of P.

Let us return once more to the committee formulation of our problem. If there are a fairly large number of committees, it is not always easy to verify that the diversity condition is satisfied. One may ask, therefore, whether it might be possible to give some simple rule for the selection of the committees insuring that distinct secretaries can always be found.

This is actually feasible. To illustrate what we have in mind, suppose that every committee has at least 5 members. Then in the graph there are at least 5 edges from each vertex in C. From a group of k committees, there would be at least $5k$ edges to vertices in P (see Figure 4.5 for $k = 4$). Now if we restrict the number of committees to which any pupil can belong to at most 5, this means that the edges from the k committees must go to at least k individuals in P, and hence the diversity condition is fulfilled.

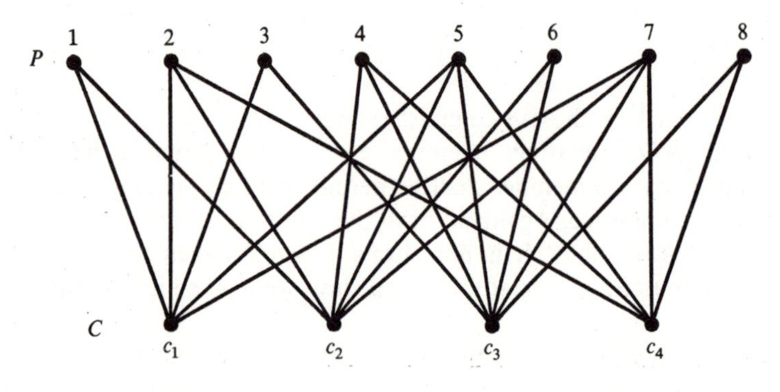

Figure 4.5

This argument is quite general, and so we may formulate this result:

suppose that each committee has at least t members; furthermore, no individual is permitted to belong to more than t committees. Then it is always possible to find a separate secretary for each committee.

In recent years the area of graph matchings has grown into a major field of study, arising in a large number of important practical applications. As well as problems involving job allocations, there are many other industrial applications involving, for example, the transportation of goods from factories to warehouses or markets.

Problem Set 4.2

1. Give an example of a set of committees which must have separate secretaries.
2. How many committees of three members can one form in a class of 12 pupils? Would it be possible to assign separate secretaries to each of them?
3. Find a matching in Figure 4.1.

4.3 Round-Robin Matchings

In all tournaments one is faced with the question of how the individual participants should be paired as the match progresses. In the case of a knock-out tournament, the problem is simple: all losers drop out in each round and one pairs the remaining winners, possibly letting one player have a bye if there are an odd number of players left.

The problem is somewhat more complicated in the case of a round-robin tournament, one of the usual kinds of chess tournament. Here every player must play against every other player, and we wish to prepare a tournament schedule in advance, giving the pairs of opponents in each round.

This situation can again be interpreted conveniently in graph terms. We suppose that there are n players, so that each of them plays $n - 1$ games with the other participants. A game is represented as before, as an edge ab connecting two players or vertices a and b. The totality of games then corresponds to the complete graph on n vertices. In Figure 4.6 the situation is presented for $n = 6$.

A round in the tournament consists in matching the players in pairs; for the moment we suppose that n is an even number, so that this may be done. In the graph the matching corresponds to a selection of $\frac{1}{2}n$ non-adjoining edges, one at each of the n vertices. In the next round we must select an entirely different set of $\frac{1}{2}n$ edges, and so on, until all games have been played. In Figure 4.6 the edges have been marked in this way: those carrying the number 1 belong to matchings in the first round, those with the number 2 to the second round, and so on.

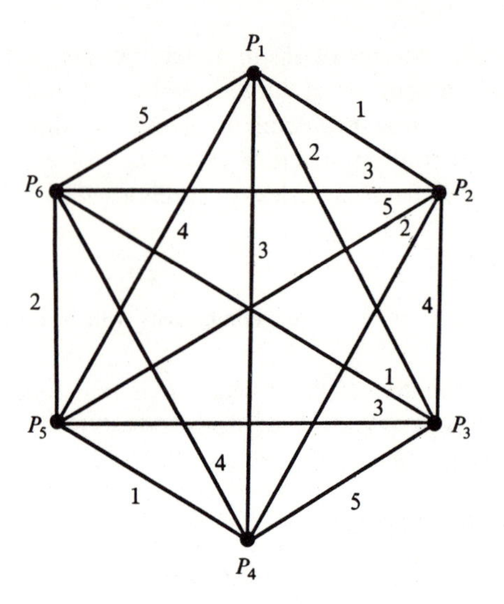

Figure 4.6

If there are a large number of players, the actual production of a schedule for the matchings in the various rounds becomes quite laborious unless there is some systematic method for doing it. Most handbooks on tournament management include extensive tables of matchings for various numbers of players, $n = 6, 8, 10, 12$, etc. For the arrangement of the schedules it is sufficient to suppose, as above, that the number of players n is even so that they may be paired; if there should be an odd number of players, we can always add a fictitious player f and then stipulate that anyone who is matched with f has a bye in that particular round.

Let us describe a simple general method for the construction of a tournament schedule for an even number of players n. We number the players $1, 2, \ldots, n$ and write these figures, in this order, in the first line of a square array. We want the next line in the array to feature the opponents of the players named in the first line during the first round of the match. Similarly, on the line below that, we want to name the opponents of players $1, 2, 3, \ldots, n$ in the second round, and so on, until all players have played each other. Clearly, our scheme should be such that all possible pairs of players encounter each other exactly once. One way of achieving this is shown in the table below. To identify the opponent of player j in the kth round, just look in the column under j, k lines below, in the array shown in Figure 4.7.

1	2	3	4	5	6	·	·	·	n
2	~~2~~1	n	$n-1$	$n-2$	$n-3$	·	·	4	3
3	4	~~3~~1	2	n	$n-1$	·	·	6	5
4	6	5	~~4~~1	3	2	n	·	8	7
·	·	·	·	·	·	·	·	·	·
·	·	·	·	·	·	·	·	·	·
·	·	·	·	·	·	·	·	·	·
$\frac{1}{2}n+1$	n	$n-1$	$n-2$	·	·	·	·	3	2
$\frac{1}{2}n+2$	3	2	n	$n-1$	·	·	·	5	4
·	·	·	·	·	·	·	·	·	·
·	·	·	·	·	·	·	·	·	·
·	·	·	·	·	·	·	·	·	·
n	$n-1$	$n-2$	·	·	·	·	·	2	1

Figure 4.7

This table is made as follows:

The first row, as already observed, enumerates the players from 1 to n; we also fill these numbers into the first column (see Figure 4.7). Now into the remaining $n-1$ places in each row we enter the numbers from 2 to n in cyclic descending order. The first $\frac{1}{2}n$ lines begin with the even numbers $2, 4, \ldots, n$, and read

$$\begin{array}{cccccccccc}
2 & n & n-1 & \cdot & \cdot & \cdot & \cdot & \cdot & 4 & 3 \\
4 & 3 & 2 & n & n-1 & \cdot & \cdot & \cdot & 6 & 5 \\
\cdot & \cdot & \cdot & \cdot & \cdot & \cdot & \cdot & \cdot & \cdot & \cdot \\
\cdot & \cdot & \cdot & \cdot & \cdot & \cdot & \cdot & \cdot & \cdot & \cdot \\
n & n-1 & n-2 & \cdot & \cdot & \cdot & \cdot & \cdot & 3 & 2
\end{array}$$

the next $\frac{1}{2}n - 1$ lines begin with the odd numbers $3, 5, \ldots, n-1$, and read

$$\begin{array}{cccccccccc}
3 & 2 & n & n-1 & \cdot & \cdot & \cdot & \cdot & 5 & 4 \\
5 & 4 & 3 & 2 & n & n-1 & \cdot & \cdot & 7 & 6 \\
\cdot & \cdot & \cdot & \cdot & \cdot & \cdot & \cdot & \cdot & \cdot & \cdot \\
\cdot & \cdot & \cdot & \cdot & \cdot & \cdot & \cdot & \cdot & \cdot & \cdot \\
n-1 & n-2 & \cdot & \cdot & \cdot & \cdot & \cdot & \cdot & 2 & n
\end{array}$$

Observe that we have failed to enter the number 1 in this table; on the other hand, since a player does not play against himself, the number at the head of any column should never be repeated within that column. Both matters are taken care of if we replace all numbers occurring in the main diagonal by 1s, as indicated. Now each player finds under his number the player he is to have as opponent in the various rounds—for instance, the player having the number 4 plays with $n - 1$ in the first round, with 2 in the second round, with 1 in the third round, 6 in the fourth round, and so on, until he finally plays with $n - 3$ in the $(n - 1)$st round.

In recent years construction methods for tournaments have grown increasingly sophisticated, and have been applied to problems increasingly remote from the game-scheduling problem discussed above.

Problem Set 4.3

1. Construct tournament tables for $n = 6, 8, 10$.
2. To prove that the preceding table is actually a usable schedule, verify the following facts:
   (a) each player has one game with every other player in the tournament;
   (b) each player has a different opponent in each round;
   (c) when player j plays player k in a round, then the table must also give j as the player opposing k.

CHAPTER FIVE

Directed Graphs

5.1 Team Competitions Re-examined

In the first chapter we used team competitions as a way of introducing graphs (Section 1.1). We joined two teams, say a and c, by an edge ac in the corresponding graph whenever these two teams had played together (see Figure 1.1). But when the various games played have been represented in this manner there is one essential fact missing: who won the game?

This deficiency can readily be remedied. Usually one draws an arrowhead on the edge ac in question. When this arrowhead points from a to c, we let it signify that team a won over team c. Suppose that we have the record of the outcomes of the various games which have been played, and add all the corresponding arrowheads in the graph in Figure 1.1. It may then appear as depicted in Figure 5.1. From this graph we read off that a won over c, but f lost to d, while b won all its games with c, e and f, and so on.

A graph G where a direction is indicated for every edge we call a *directed graph* (often abbreviated to *digraph*). It may be intended, as we indicated, to represent the results of a competition between teams or individuals; on the other hand, any directed graph can be conceived of as a representation of a competition.

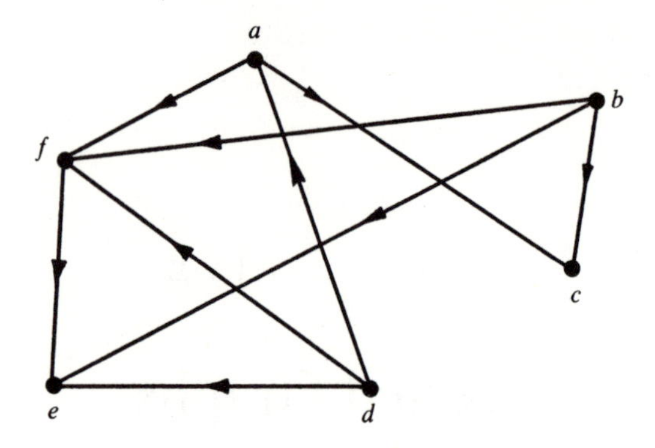

Figure 5.1

We have passed over one point in silence: what happens when a game ends in a draw? On the whole, draws are a nuisance for the score keeping in any kind of a tournament. Often the rules are so formulated that draws cannot occur, as, for instance, in tennis or squash. In other games, such as golf and soccer, the players and teams play extra rounds to avoid an undecided match. But if draws are unavoidable, then we may take them into account in the graph by letting the corresponding edges remain undirected. Then we obtain a *mixed graph* in which some edges are directed and others undirected. This type of graph occurs also in other problems, as we shall see.

As an illustration of a mixed graph, we may take Figure 5.2 in which the four teams a, b, c, and d all have played a game with one another: a has won over b and c, and drawn with d; b has lost to a, drawn with c, and won over d; c has lost to a, won over d, and has drawn with b; d has drawn with a, and lost to b and c.

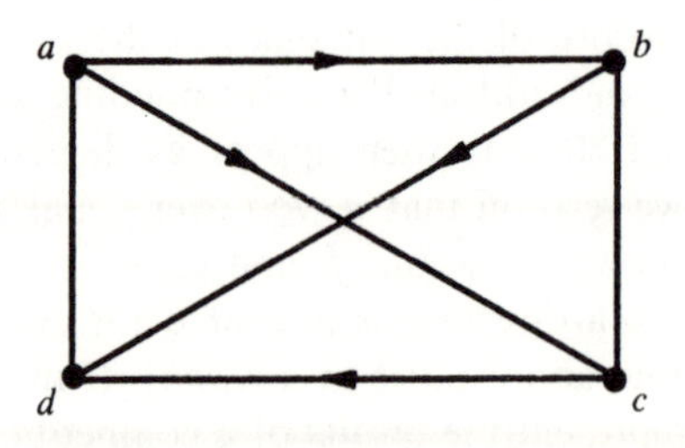

Figure 5.2

5.2 The Problems of One-Way Traffic

The map of any network of roads or streets furnishes us with a somewhat special, yet illustrative, example of a graph. But an up-to-date city plan should show not only the relative locations of the streets and their intersections; it should also give information about which streets have a two-way traffic flow and which streets have one-way traffic, together with the direction in the latter case. Clearly, we are again faced with a directed graph, or rather with a mixed graph if not all streets have one-way directions (Figure 5.3). We can make the whole graph directed by a device often used in graph theory—namely, by replacing an undirected edge by two directed edges, one in each direction, between the same two vertices (see Figure 5.4).

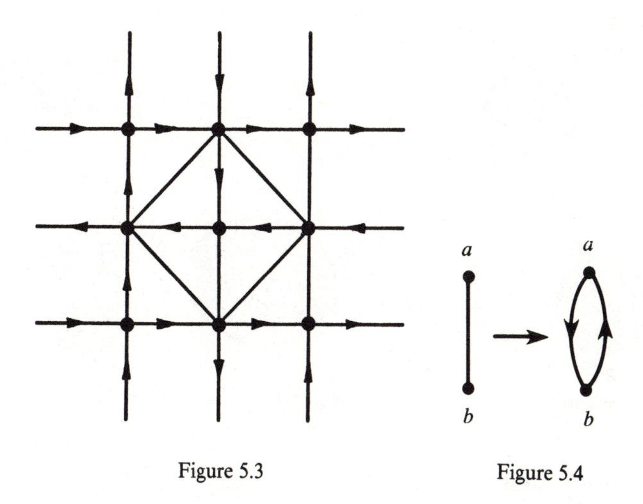

Figure 5.3 Figure 5.4

One-way traffic in a city poses questions which are of interest also for general graphs. Suppose that the police department in a town decides to introduce a new traffic plan so radical that every street is made one-way. This would certainly create an awful outcry from the citizenry if it were discovered that one could not always drive legally from one place to another. This brings up the problem: *when is it possible to direct the streets in such a manner that one can drive from any point to any other along the prescribed directions*? In the more general graph formulation the corresponding problem reads: *when can the edges of a graph G be given directions in such a way that there is a directed path from any vertex to any other*?

It is clear that the graph must be connected. But there are other conditions which must be satisfied. A graph edge $e = ab$ is called a *bridge* when there is no way of getting from a to b, or vice versa, except through e. A bridge e divides the vertices in the graph G into two sets—namely, those vertices one can reach from a without traversing e, and those one can reach from b without traversing e. This corresponds to a separation of the graph G into two parts G_1 and G_2, connected only by the edge e (see Figure 5.5).

On a city map, a bridge is a single connection between two separate parts of the town; perhaps it may be a single bridge over a river, or a single railroad underpass. Clearly, if such a connection were made into a one-way street, no vehicle could leave one part of the town to get to the other.

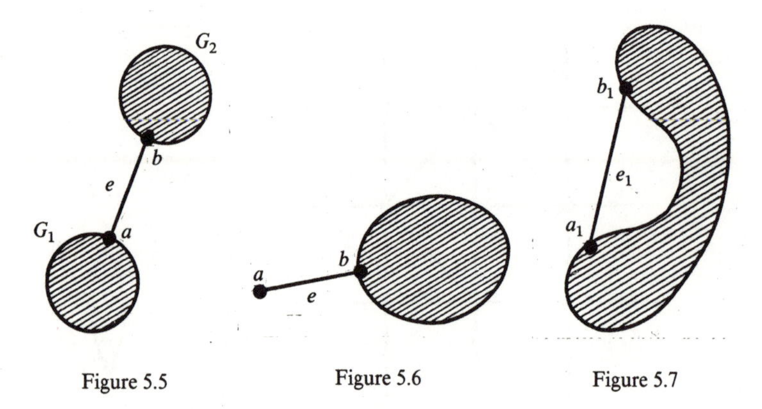

Figure 5.5 Figure 5.6 Figure 5.7

Earlier (Section 3.1), we said that an edge $e = ab$ is a *terminal edge* if at one of the endpoints (for instance, at a) there are no other edges of the graph (see Figure 5.6). Such an edge must also be regarded as a bridge, since there is no way of getting from a to b except through e. We may consider that in this case the graph G_1 in Figure 5.5 has shrunk to a single vertex a. On a street map, a terminal edge corresponds to a dead-end street; it cannot be made one-way without blocking the access to a, or the exit from a.

If $e_1 = a_1b_1$ is an edge which is not a bridge, then there must be some other way from a_1 to b_1 which does not pass through e_1 (see Figure 5.7). For this reason, such an edge e_1 is called a *cycle edge*. There are therefore two types of edges in the graph, the cycle edges and the bridges.

We are now ready to prove the following result:

THEOREM 5.1. *If G is an undirected connected graph, then we can always direct the cycle edges of G and leave the bridges undirected, so that there is a directed path from any given vertex to any other.*

In terms of a city map, this may be expressed as follows: if we leave the single bridges and dead-end streets as two-way passages, all other streets can be made one-way in such a manner that the traffic is assured satisfactory connections everywhere.

We can prove the theorem by giving a method for directing the graph edges suitably. We begin by taking some arbitrary edge $e = ab$ in G. If e is a bridge, it remains two-way, and so we can get to b from a and vice versa along e (Figure 5.8). On the other hand, if e is a cycle edge on a cycle C all edges on C may be directed in circular fashion; evidently we can always get from one vertex to another on C by following the edge directions (Figure 5.9).

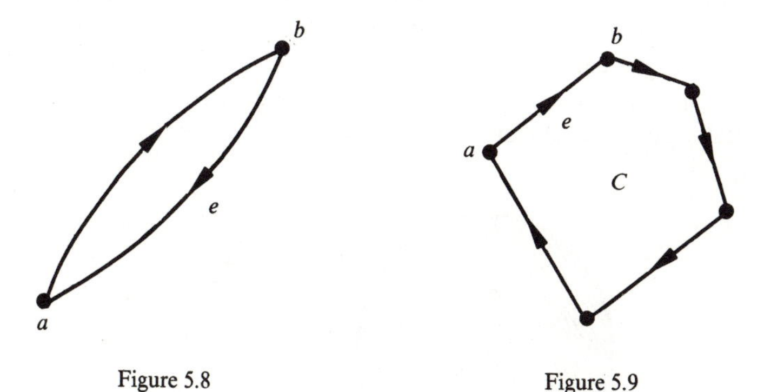

Figure 5.8

Figure 5.9

We have made a small beginning on directing the graph, and this can be steadily enlarged. Suppose that we have directed a certain part H of the graph G in question, so that the theorem holds for H. Since G is connected, if H is not the whole graph then there must be some edge $e = ab$ touching H; that is, e does not belong to H, but has one of its endpoints (say, a) on H.

If e is a bridge, we agreed that it must remain two-way. Therefore, from any vertex x in H, we can proceed by a directed path P to a, and then through e to b. Conversely, we can go from b through e to a, and then by some directed path Q from a to x (Figure 5.10). Thus we can add e to H, and have a larger part of G which is properly directed.

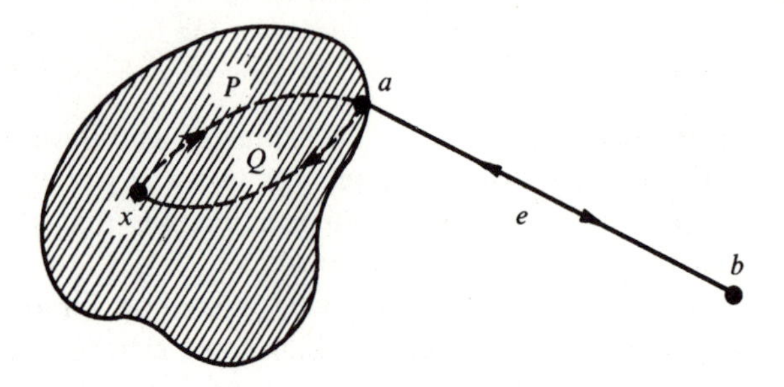

Figure 5.10

In the other case, $e = ab$ is a cycle edge and so lies on a cycle C. We proceed on C from a to b, and further to the next vertex d of C which lies on H (Figure 5.11). We may have $a = d$. We put a direction on the edges in C from a to b and further on to d, and add these edges to H. Then we can go by a directed path P in H from any vertex x in H to the vertex a, and then along C to any vertex y on C. Conversely, we can proceed from y along C to d, and by a directed path Q in H from d to x. By enlarging H repeatedly in this manner, we finally get all edges directed as desired.

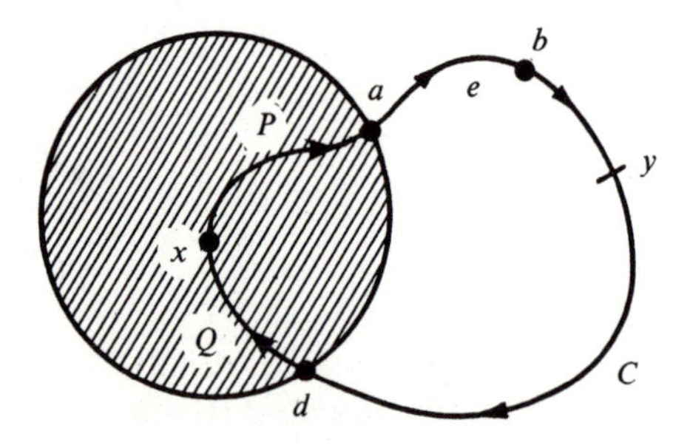

Figure 5.11

Creating one-way streets is certainly an effective method of expediting the flow of traffic, but it may also pose real puzzles for a driver unfamiliar with the town who wants to find his way around in a car. In the case of two-way street traffic, we can use the procedure outlined in Section 2.4 to pass through every edge of an undirected graph. But this manner of groping around depends for its success essentially upon the fact that we can pass through each edge in both directions. When the streets are one-way, wholly or in part, the situation becomes much more complicated, even when we suppose, as we must, that there is always at least one directed path from any point to any other.

The general problem is: *how can we proceed systematically along the given directions of the graph so that we will eventually pass through every edge*? The crux of the matter is to have a good memory; better still, we can draw a graph sketch of the streets or edges as we proceed.

We start at some point a_0 through one of the streets issuing from it. Whenever we pass a street intersection, we mark on our sketch which street we came from and which new street we take, and we indicate also the other streets at this corner, together with their directions; these will be explored later. After a while, we must return to an intersection a_1 which has been visited before (Figure 5.12).

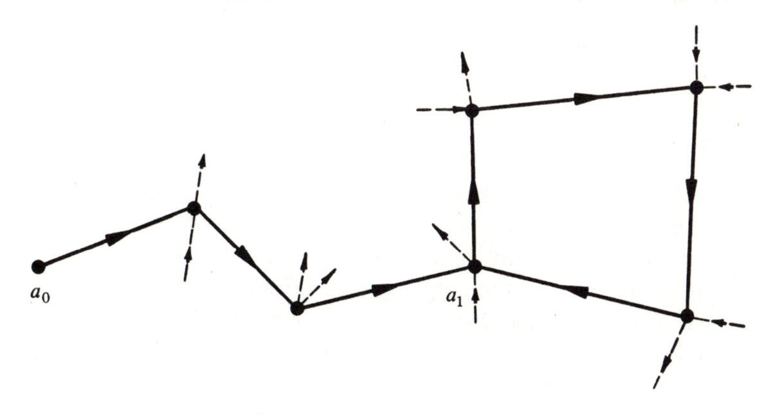

Figure 5.12

We look at the part of the map prepared after the departure from a_1, to see which side streets have not been traversed on this section. We select one of them, and proceed to it and then through it. By continuing this process, we cover more and more streets; many of them will be covered several times, but there will also always be new ones until the whole graph has been traversed. Until this happens, we can never be stymied for lack of new streets to pass. Suppose, for example, that at some stage we were at a vertex a_i and there was still a street cd that we had not traversed. By our assumption, there is a directed path from a_i to c, and this would at some point depart from the paths we had already covered.

Problem Set 5.2

1. Use the preceding method to direct the edges in the graphs in Figure 1.14 and Figure 3.4.
2. Do the same for the street maps in Figure 5.13 and Figure 5.14.

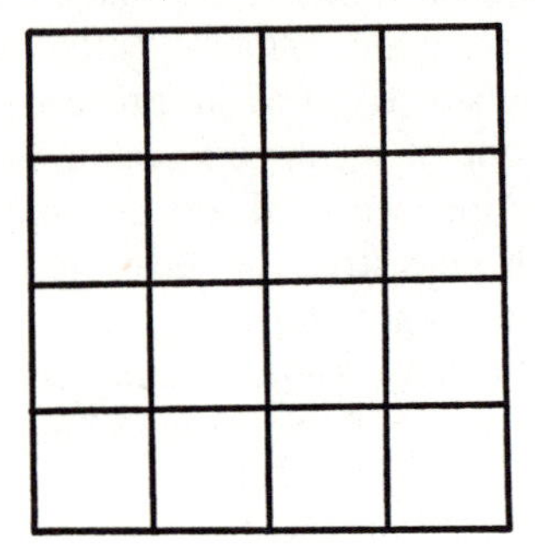

Figure 5.13

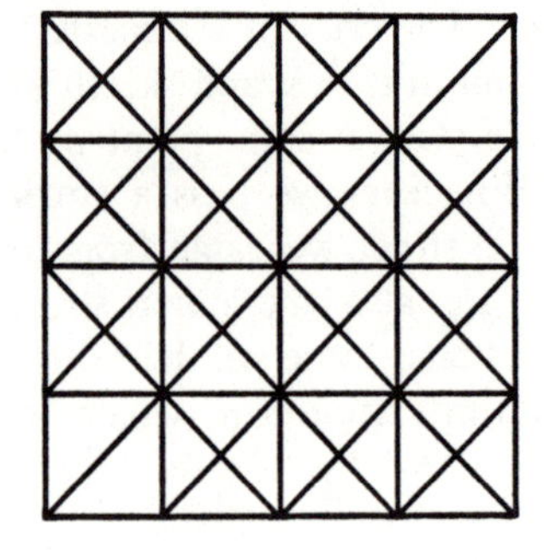

Figure 5.14

5.3 Degrees

In Section 1.6 we considered the number of edges of an undirected graph; the degree $\deg(a)$ at a vertex a was the number of edges having a as one endpoint. In a directed graph there are two types of edges at each vertex a: the outgoing edges from a, and the incoming edges to a. Correspondingly, we have two degrees—the number $\mathrm{outdeg}(a)$ of outgoing edges and the number $\mathrm{indeg}(a)$ of incoming edges. In Figure 5.15 we have

$$\mathrm{outdeg}(a) = 3, \quad \mathrm{indeg}(a) = 2.$$

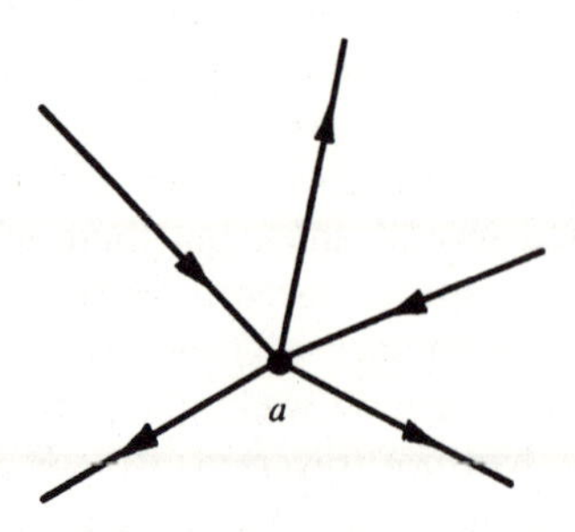

Figure 5.15

Each directed edge $e = ab$ has one initial vertex a and one terminal vertex b. We can therefore obtain the total number m of edges either by counting the number of outgoing edges at each vertex, or by counting the total number of incoming edges. This means that for a directed graph with n vertices

$$a_1, a_2, \ldots, a_n$$

the number m of edges is given by either one of the expressions

$$\begin{aligned} m &= \text{outdeg}(a_1) + \cdots + \text{outdeg}(a_n) \\ &= \text{indeg}(a_1) + \cdots + \text{indeg}(a_n). \end{aligned}$$

As an example we may take the directed graph in Figure 5.1. Here

$$\text{outdeg}(b) = \text{outdeg}(d) = 3, \quad \text{outdeg}(a) = 2, \quad \text{outdeg}(f) = 1,$$
$$\text{outdeg}(c) = \text{outdeg}(e) = 0,$$

while

$$\text{indeg}(e) = \text{indeg}(f) = 3, \quad \text{indeg}(c) = 2, \quad \text{indeg}(a) = 1,$$
$$\text{indeg}(b) = \text{indeg}(d) = 0,$$

and in either case the total is 9.

There are various types of directed graphs in which the degrees have special properties. Let us mention a few. A directed graph is called *regular of degree r* if all degrees have the same value r:

$$\text{outdeg}(a) = \text{indeg}(a) = r,$$

for each vertex a. A simple example is a cycle (Figure 5.16); here

$$\text{outdeg}(a) = \text{indeg}(a) = 1$$

for every vertex a, so that the directed graph is regular of degree 1.

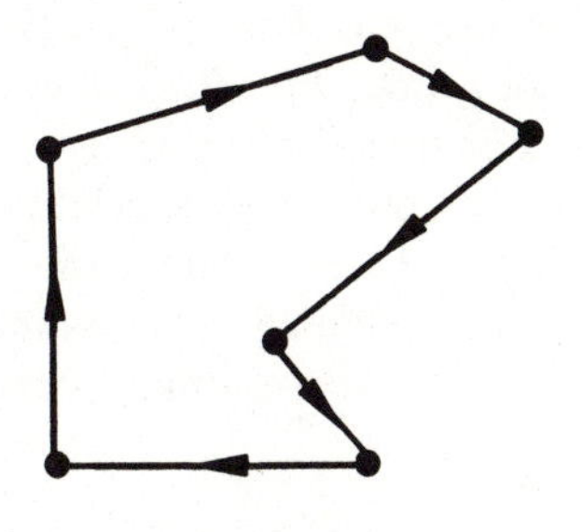

Figure 5.16

As another example, let us take the directed graph of a round-robin tournament in which each team is scheduled to play with every other. Suppose first that there are an even number of competing teams, so that there are no byes. Then, after k rounds, each team will have played k others. In the corresponding directed graph, there must therefore be a total of k edges at each vertex—some incoming, denoting losses, and the others outgoing, denoting victories (excluding the possibility of draws). Consequently, the degrees of our directed graph must satisfy the conditions

$$\text{outdeg}(a) + \text{indeg}(a) = k,$$

at each vertex a.

If the number of teams is odd, then in each round there will be one team that does not play. For the teams which have played in every round the above relation remains satisfied, while for the k teams b which have had a bye it becomes

$$\text{outdeg}(b) + \text{indeg}(b) = k - 1.$$

Problem Set 5.3

1. Draw the directed graphs of some tournaments of 5 teams after 2 and 3 rounds.
2. How must the above degree relations be modified when draws may occur?
3. Draw regular directed graphs of degree 2, for $n = 5, 6, 7$ and 8 vertices.

5.4 Genetic Graphs

When you draw your family tree, you may make use of a directed graph to illustrate the family relationships. A directed edge ab is drawn from a member a to another member b to indicate that b is a child of a. Biologists use this kind of diagram systematically to describe the outcome of genetic breeding experiments, denoting by a directed edge ab that b is the offspring of a.

Such genetic graphs have some very special properties which come to light almost immediately. One of them is the consequence of sexual reproduction. Since each individual has two parents, one male and the other female, there are just two incoming edges to each vertex b; that is,

$$\text{indeg}(b) = 2.$$

Thus the basic figure in a genetic graph consists of two edges, as in Figure 5.17, where b is the offspring of two individuals, the male m and the female f.

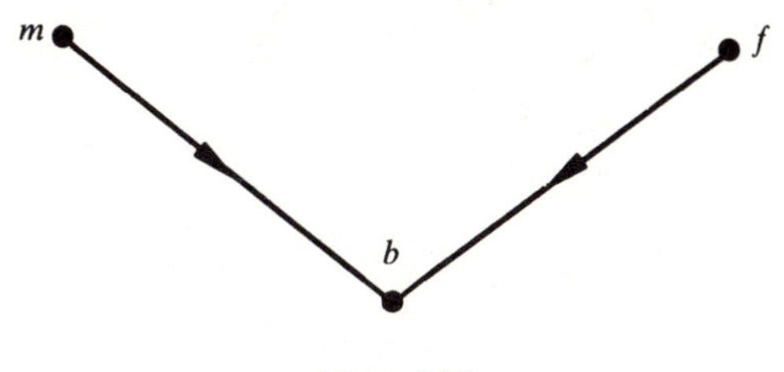

Figure 5.17

We may remark at this point that family trees, like ordinary trees, do not extend into the heavens. Since our knowledge has limits, we always reach a point at which neither parent, or only one parent, is known. Thus we might properly replace the above equality by the inequality

$$\operatorname{indeg}(b) \leqslant 2.$$

Any family relationship is expressed by a configuration in the genetic graph. So, for instance, Figure 5.18 informs us that the individuals b_1, b_2, b_3 are brothers or sisters, since they are all the children of the same parents m and f.

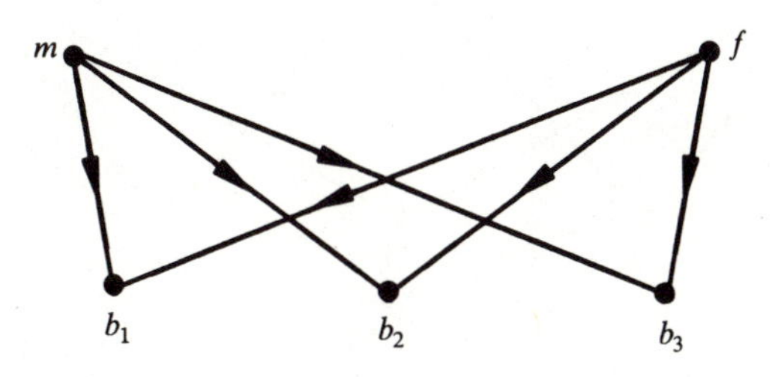

Figure 5.18

Similarly Figure 5.19 tells us that b_4 is a half-brother or half-sister of the children b_1, b_2, b_3, since they have the same mother f, but different fathers, m and m_1.

We now come to a point of interest. Suppose that we have a directed graph with at most two incoming edges at each vertex. We may then ask: is it possible to introduce two sexes in the graph—in

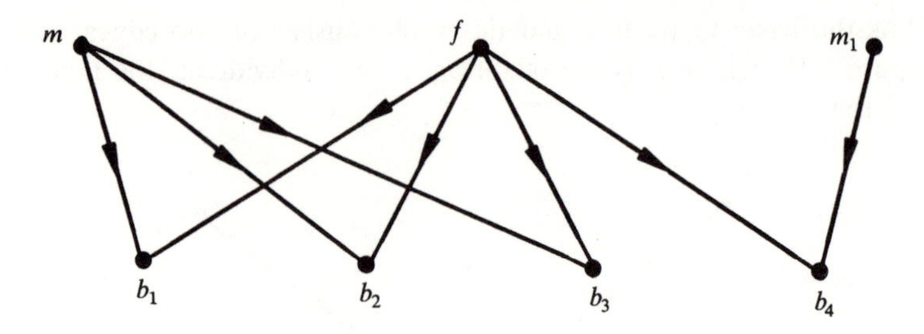

Figure 5.19

other words, to divide the vertices or individuals into two classes, male m and female f, such that the two incoming edges to any vertex always originate one from each group as in Figure 5.17?

An example (Figure 5.20) shows that this is not always possible. Suppose that we take a_1 to be male. Since he has a child b_1 with a_2, the latter must be female. Since a_2 and a_3 have a child b_3, a_3 is also a male. But this contradicts the fact that according to the directed graph, b_2 is a child of a_1 and a_3, both males. A similar contradictory situation results if one supposes that a_1 is female.

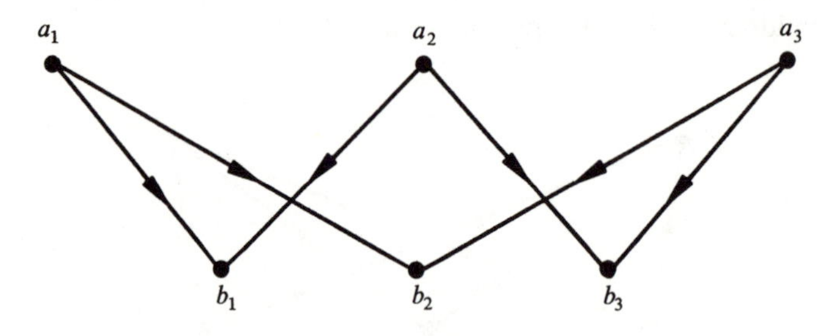

Figure 5.20

To analyze the cause of this trouble, let us start with some vertex a_1, which we shall call male. If a_1 has a child b_1 with another parent a_2, then a_2 must be female. If a_2 also has a child b_2 with another individual a_3, then a_3 is male, and so on. In this manner we obtain an alternating sequence of males and females

$$a_1, a_2, \ldots, a_n.$$

In our directed graph they are connected by a sequence of edges, alternately having opposite directions,

$$a_1b_1,\ a_2b_1,\ a_2b_2,\ldots,\ a_nb_{n-1},$$

as indicated in Figure 5.21.

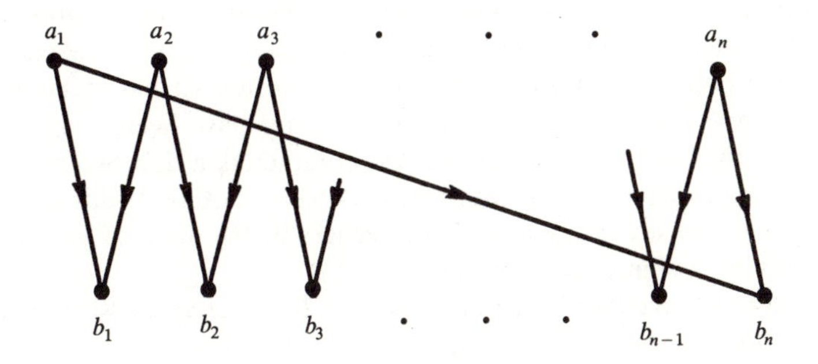

Figure 5.21

Suppose now that also a_1 and a_n have a child b_n as in Figure 5.21. This results in a *circular alternating path*

$$a_1b_1,\ a_2b_1,\ldots,\ a_nb_n,\ a_1b_n,$$

running back into itself. In order that this be compatible with the alternating sex determination in the sequence $a_1, a_2,\ldots, a_n$, the vertices a_1 and a_n must have opposite sex—that is, n must be an even number. This leads to the following result:

Sex condition: Let G be a directed graph in which indeg$(b) \leqslant 2$, *for each vertex b. If a division of the vertices of G into two sex groups is possible, then each sequence $a_1, a_2,\ldots, a_n$ of parents in a circular alternating path must have an even number n of terms.*

An equivalent statement of this condition is that, in any circular alternating path $a_1b_1, a_2b_1,\ldots, a_nb_n, a_1b_n$, the number of edges is divisible by 4.

If the sex condition is satisfied, we can assign a sex to each vertex in the directed graph. We begin with a vertex a_1 and assign a sex to it arbitrarily. If a_1 has no children, or if these children have no other known parents, this ends the sex consequences of a_1. But if a_1 has a child in common with some individual a_2, then one forms all alternating paths from a_1. This determines uniquely the sex of each parent $a_1, a_2,\ldots, a_n$; for, if there were an alternating path from a_1 to a_n making a_n male, and another path from a_1 making a_n female, then we could proceed from a_1 to a_n on the first path, and return to a_1 on the second. But this would produce a circular alternating path with an odd number of vertices, contrary to the requirement of our sex condition.

In this first step, not all vertices will have been given a sex. In the next step, we select some vertex a_1', not related to a_1 by an alternating path; to a_1' we assign a sex character arbitrarily, and continue as before; then a third vertex a_1'' unrelated to a_1' and a_1 is taken as a starting point, and so on, until the whole vertex set has received its characters. From the way in which the sexes have been assigned, it follows that we can never run into the contradiction that two males or two females have a child together. It is worth noting in this connection that the sex characters of the vertices in the directed graph can usually be assigned in many ways.

After we have discovered the condition for a suitable assignment of sexes to the vertices, it is natural to ask whether a directed graph satisfying this condition can actually be realized as the image of a breeding experiment. We find readily that one further restriction must be imposed on the directed graph.

Suppose that we have a sequence of individual organisms

$$d_1, d_2, \ldots, d_n,$$

each an offspring of the preceding one. In the directed graph, this corresponds to a directed path (Figure 5.22). The births of these individuals $d_1, d_2, \ldots, d_n$, must follow in the same order in time. Consequently, it cannot happen that d_n becomes the parent of d_1 (Figure 5.22)—that is, the directed graph cannot contain any cyclic directed path; such directed graphs are called *acyclic*.

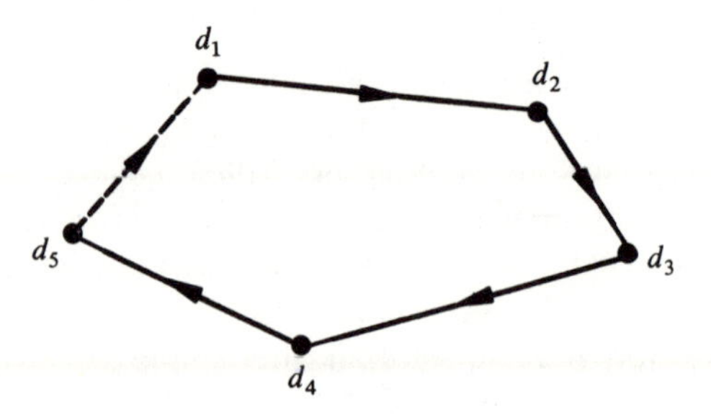

Figure 5.22

We have established three necessary conditions for a directed graph to be interpreted as a genetic experiment:

(i) $\text{indeg}(a) \leqslant 2$, for each vertex a: no vertex has more than two incoming edges;
(ii) each circular alternating path has a number of edges divisible by 4;
(iii) the directed graph is acyclic.

Conversely, when these conditions are satisfied and sex characters have been suitably assigned, the whole directed graph may be considered as depicting what happens in a genetic experiment; an edge $a_0 b$ signifies that b is the offspring of a_0 from a mating with some other (possibly unknown) individual a_1 of the opposite sex.

Our three conditions may therefore be considered to be the rules for a genetic scheme in general. In more everyday language they are the self-evident axioms:

(i) each individual has at most two parents;
(ii) the parents belong to opposite sexes;
(iii) no individual is his own ancestor.

To conclude, let us make one further observation about genetic graphs. We have considered our directed graph as representing the results of an arbitrary genetic experiment in which matings may be induced whenever biologically feasible. In human society the situation is otherwise; well-recognized taboos exclude a variety of configurations from our family trees. For instance, since no individual may marry his own sister or brother, there can be no configurations in our directed graph of the form given in Figure 5.23. Since no individual may marry one of his parents, there are no configurations of the type represented in Figure 5.24.

Figure 5.23

Figure 5.24

Problem Set 5.4

1. Draw the directed graph which indicates that two individuals are: (a) cousins; (b) aunt and niece.
2. Assign sex characters to the vertices in the graph in Figure 5.25 in all possible ways.
3. Draw the configurations which are excluded from a family tree by the taboos against marriages between: (a) half-brother and half-sister; (b) grandparent and grandchild; (c) uncle and niece.

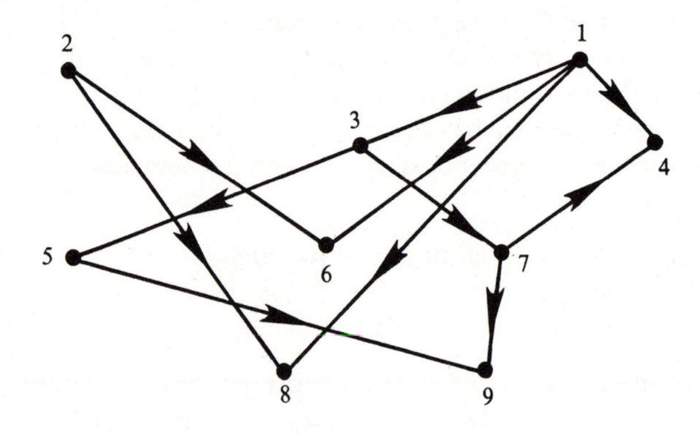

Figure 5.25

5.5 Finding the Shortest Route

Consider the road network shown in Figure 5.26; the numbers next to the edges are the times taken to travel along the corresponding

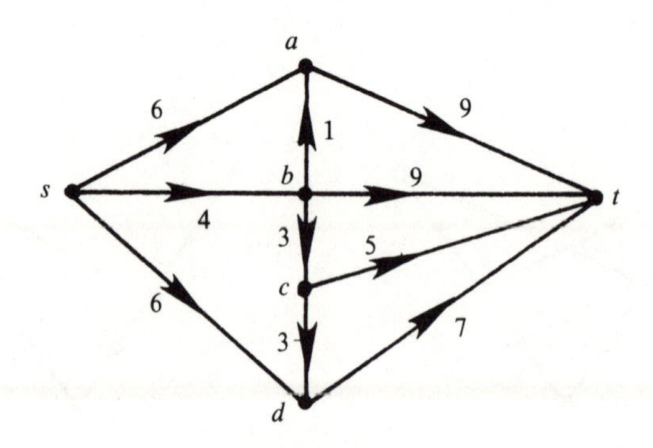

Figure 5.26

stretch of road. Can you plan a route from city s to city t which covers the journey in the least possible time?

In this section we shall describe a simple method for finding the shortest route between any two given cities in a road network. This method is very efficient and can be used for any network, however complicated. The basic idea is to move across the network from left to right, determining the shortest time from city s to each of the intermediate cities as we go. At each intermediate stage we look at all those vertices reachable by an edge from the current vertex and assign to each one a temporary label (which we write in a circle), representing the shortest time taken to reach it by paths already considered. Eventually, each vertex receives a permanent label (which we write in a square), representing the shortest time from city s to that city. Our aim is to find the permanent label assigned to city t. The method will become clear as you work through the following example.

We start by assigning city s the permanent label 0 (since it takes no time to go from s to s), and look at all cities reachable directly from s. These cities are a, b and d, and we assign to them temporary labels of 6, 4 and 6, corresponding to the times taken to reach them directly from s.

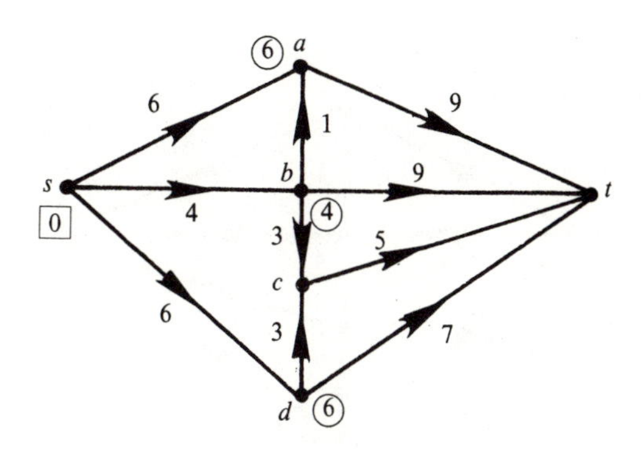

Figure 5.27

The next step is to take the smallest temporary label and make it permanent. This occurs at city b, and shows that the shortest time to travel from city s to city b is 4. We now look at all cities reachable directly from b (that is, cities a and c), and assign to each one a temporary label equal to the label at b plus the time taken to go from b to that vertex—unless that vertex already has a smaller label. In this case, we assign to city a the new label $4 + 1 = 5$, and to city c the label $4 + 3 = 7$.

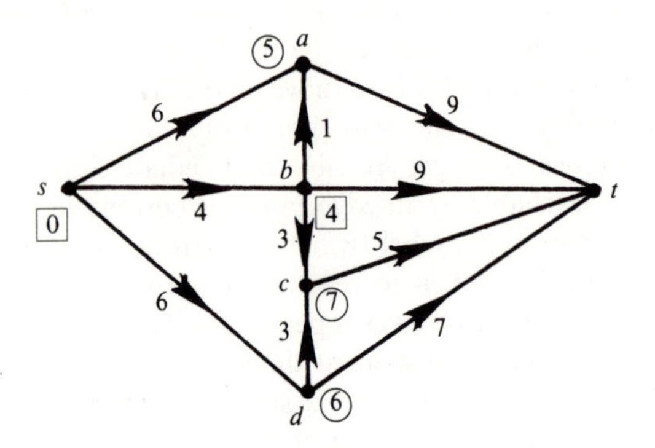

Figure 5.28

The smallest temporary label is now 5 at a, so we assign to city a the permanent label 5. We now look at all cities reachable directly from a (that is, city t), and assign to it the temporary label $5 + 9 = 14$.

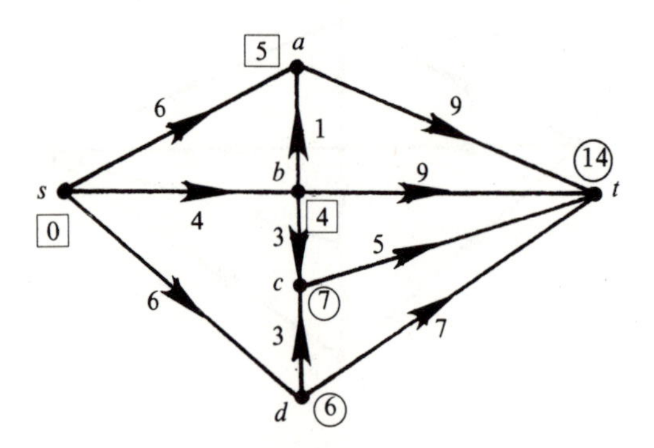

Figure 5.29

The smallest temporary label is now 6 at d, so we assign to city d the permanent label 6. We now look at all cities reachable directly from d (that is, cities c and t), and assign to city t the new label $6 + 7 = 13$. We leave the label at c unchanged, since $6 + 3$ is larger than its present label 7.

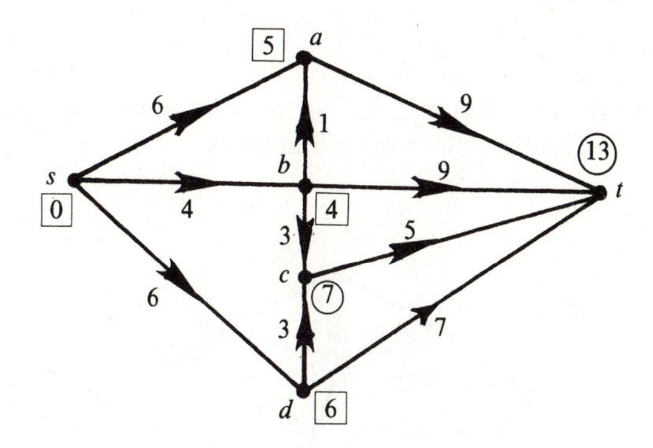

Figure 5.30

The smallest temporary label is now 7 at c, so we assign to city c the permanent label 7. We now look at all cities reachable directly from c (that is, city t), and assign to it the new label $7 + 5 = 12$. The only remaining temporary label is 12 at t, so we assign to city t the permanent label 12. Thus, *the shortest time to reach city t is 12*.

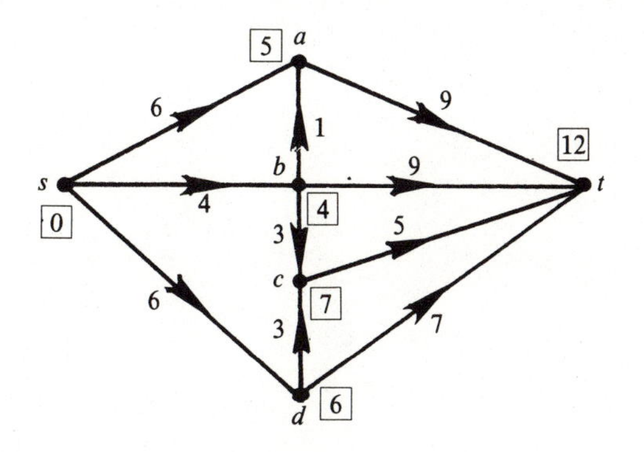

Figure 5.31

We can find the shortest route from city s to city t by working backwards from t, as follows. Since

$$(\text{the label at } t) - (\text{the label at } c) = \text{the time } ct,$$

we include the edge ct. Similarly,

$$(\text{the label at } c) - (\text{the label at } b) = \text{the time } bc,$$

and $$(\text{the label at } b) - (\text{the label at } s) = \text{the time } sb,$$

so we include the edges bc and sb. Thus, *the shortest route from city s to city t is sbct*.

Problem Set 5.5

1. Use the above method to find the shortest time and the shortest route from city s to city t in the following road network:

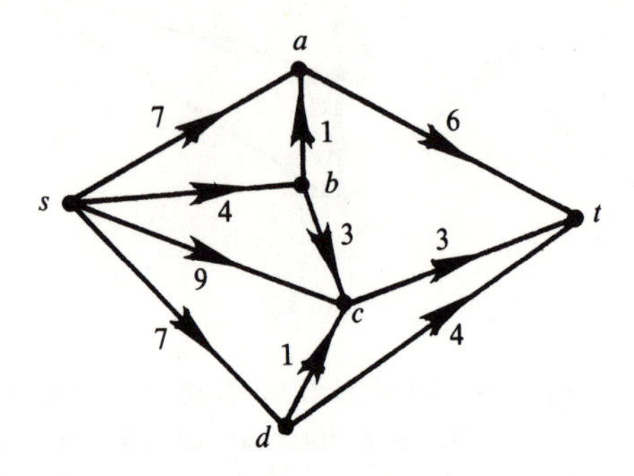

Figure 5.32

CHAPTER SIX

Questions Concerning Games and Puzzles

6.1 Puzzles and Directed Graphs

Previously (Section 2.6) we explained how puzzle problems can be formulated in terms of graphs. The vertices of the graph correspond to the positions in a puzzle; the edges of the graph correspond to the possible moves from one position to another. The solution of the puzzle consists in finding a path from a given initial position to one (or possibly more) terminal or winning positions.

In dealing with these puzzles, we used undirected graphs. This was based upon the tacit assumption that the moves can be made both ways from one position to another. Such a procedure is permissible for the puzzles of the ferryman, the three jealous husbands, and the moves of the knight on the chessboard.

But for many puzzles the moves can only be made in one direction, and in this case we are compelled to use directed graphs in the representation. If some moves can be made in both directions, we can include an edge for each direction, or we can use a mixed graph in which these edges are undirected. To solve the puzzle, we must find a directed path from the initial position in the graph to the desired terminal position.

We shall illustrate these remarks by considering an ancient and familiar puzzle. We have three jugs A, B, C, with capacities $8, 5, 3$

quarts, respectively. The jug A is filled with wine, and we wish to divide the wine into two equal parts by pouring it from one container to another—that is, without using any measuring devices other than these jugs.

We shall use the following scheme for solving this puzzle graphically. To every distribution of wine in jugs B and C, we assign the pair of numbers (b, c), b denoting the amount in jug B, and c the amount in jug C. Initially (b, c) has the value $(0, 0)$, and the desired terminal distribution is $(4, 0)$; jugs A and B would then contain equal amounts of the wine, and jug C would be empty.

Since, to each pair of real numbers (b, c) we may assign a point with coordinates (b, c) in a coordinate plane, we may think of all possible distributions as points, and these will be the vertices of our directed graph. It is clear from the statement of the problem that we cannot accurately pour fractions of quarts into jugs B and C; so b can only take the values $0, 1, 2, 3, 4, 5$, and c the values $0, 1, 2, 3$. This means that there are $6 \times 4 = 24$ possible distinct pairs (b, c)—that is, 24 vertices in our directed graph. Whenever it is possible to change a known distribution (b_0, c_0) to a new distribution (b_1, c_1) by pouring wine in accurate amounts, we connect the vertex (b_0, c_0) to the vertex (b_1, c_1) by a directed edge. In our example, the edges ia_1 and ia_2 (Figure 6.1) lead away from the initial vertex $i = (0, 0)$. From a_1, it is possible to reach the distributions $a_{11} = (5, 3)$ and $a_{12} = (2, 3)$; and from a_2, we can reach $a_{21} = (3,0)$ and $a_{22} = (5, 3)$. Next, we might list all vertices attainable from these by the rules of our game, and

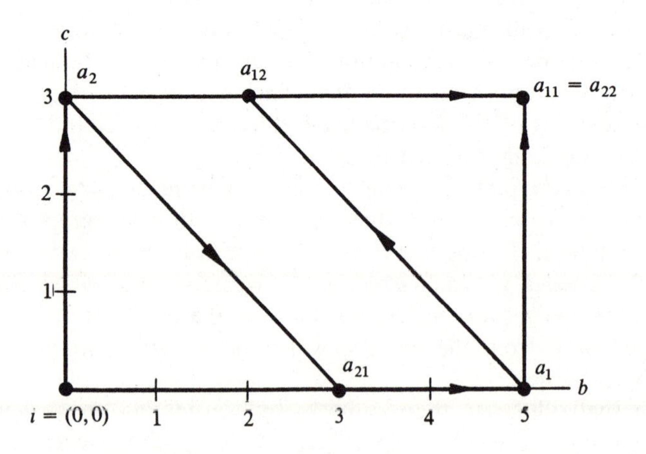

Figure 6.1

continue in this way. If the terminal vertex $t = (4, 0)$ can be reached at all, this method must eventually lead to it. In fact, there may be many paths from i to t, and we may want to select the 'best' solution (in the sense of having to pour wine as few times as possible) by determining a path from i to t with as few edges as possible.

While from a certain initial vertex i many paths may lead to a given vertex t, there are, in general, other vertices u that cannot be reached from i by the prescribed rules. In our example, these unattainable vertices u are

$$(1,1),\quad (1,2),\quad (2,1),\quad (2,2),\quad (3,1),\quad (3,2),\quad (4,1),\quad (4,2).$$

In other words, of the 24 vertices initially included as 'possible' distributions, 8 cannot be attained, and only 16 turn out to be possible.

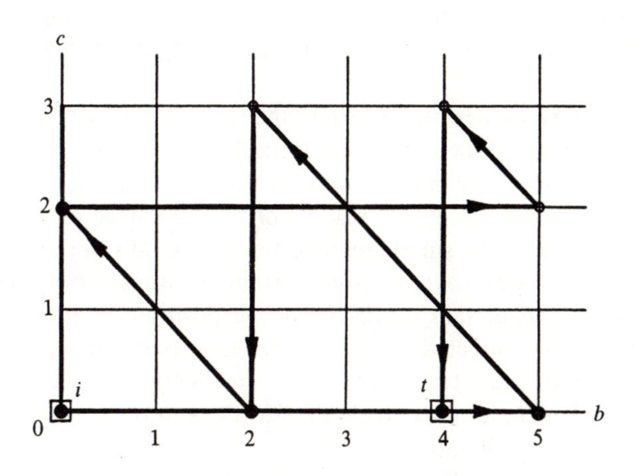

Figure 6.2

Actually, our solution of this jug puzzle (Figure 6.2) is not a very good illustration of the directed path method because all pourings can be reversed, as you may check quite readily. However, if we had started with a different distribution, say one quart of wine in each of jugs B and C (that is, from the vertex $(1, 1)$), all edges would have been directed away from $(1, 1)$.

Problem Set 6.1

1. Plot the unattainable vertices in Figure 6.2. What does their location indicate?

2. Show that we can never pass from any unattainable vertex to another by repeated pourings.
3. Solve the same problems for some other sizes of the jugs—for instance, $A = 12$, $B = 7$, $C = 4$.

6.2 The Theory of Games

A puzzle is a solitary struggle against the difficulties inherent in a problem, while in a game one plays against an opponent. Since we have entered upon the field of entertainment and discussed some puzzles, let us also make some general observations about games. The theory of two-person games has recently become an important field of mathematical research. It has applications to many problems of a practical nature; we may mention engineering and economics, where the theory of games is used to solve questions concerning the most effective or most economical way of performing certain complicated tasks.

Here we have no occasion to study the general theory of games. We shall indicate only how our directed graphs may be used to illustrate the principles of those games in which the moves do not depend on chance. Chess is an excellent example of what we have in mind; so is checkers, and even the simple-minded tic-tac-toe falls in this category.

In these games we have, as usual, certain positions corresponding to the vertices, and certain moves from one position to another corresponding to the directed edges. But in each position we must also know which of the two players Alf and Betty has the move. It therefore seems natural to divide the positions into two groups, called A and B, so that the moves are represented by directed edges from A to B or from B to A. The same position may possibly occur both in A and B.

The play then consists in Alf moving to a new position along an edge to B, and Betty back to A. We could say that this is a game with a single piece which is moved along the directed edges back and forth between the two sets. In each position the players, presumably, know what they want to do. Thus we could even do away with the players if they had recorded in a book of strategy what moves they would select under given conditions. As a consequence, the whole game is determined when we know its graph—that is, the moves permitted in the game and the strategy of the players.

To win a game, Alf must move from an initial position along a directed path, in part determined by Betty, to some winning position

w_A in B; similarly, in order for Betty to win, her last move must be to a winning position w_B in A.

A draw or undecided game may occur in two ways. There may be certain end positions in A or B which are called *draws*, and there are no further edges from them. A stalemate in chess is a good example. Another type of draw occurs when the game can go on indefinitely, usually by a repetition of some cycle of moves. To avoid this, we may stipulate that the game is over after a certain number of repetitions of the same moves. In chess, the game is drawn after the same moves have been repeated in sequence three times; this game is also limited by the rule that in 50 moves a pawn must have been moved.

We now have a perfectly good picture of a game. There remains only one essential point: *when can Alf (or Betty) play in such a way so as to be certain of winning*? This is a question which may also be answered from the graph. The best way to handle it is to reason backward from the final winning positions for Alf. Since these positions are losing for Betty, we indicate this by denoting the set of them by $L_0(B)$. In A, there are some positions with edges or moves into $L_0(B)$. From such a position, Alf can win in a single move; we call this set $W_1(A)$. Next there will be a set $L_1(B)$ of positions, either in $L_0(B)$ or having only moves to positions in $W_1(A)$. When Betty has such a position, she has either already lost or cannot avoid losing in one move. We can, therefore, just as well consider the larger set $L_1(B)$ as the set of lost positions for Betty (Figure 6.3).

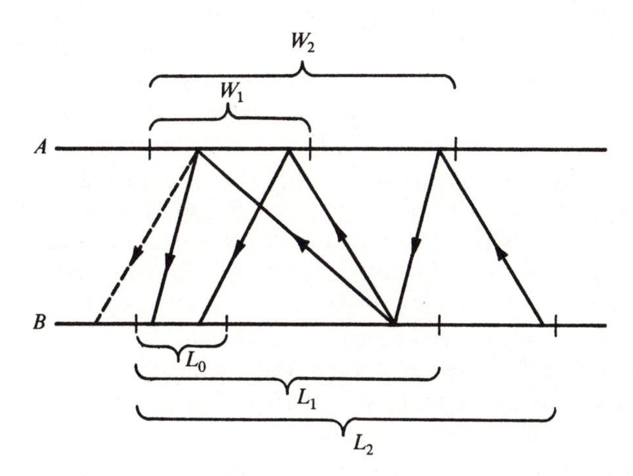

Figure 6.3

We can continue this process of enlarging the losing set for Betty. There will be a certain set $W_2(A) \supseteq W_1(A)$ of vertices in A with edges to $L_1(B)$. From such positions, Alf can win in at most 2 moves. Furthermore, there will be a certain set $L_2(B)$ of positions from which all edges go to $W_2(A)$ and Betty cannot avoid losing in at most 2 moves. When repeated, this leads to two sequences of sets in A and B, respectively:

$$W_1(A) \subseteq W_2(A) \subseteq \cdots,$$
$$L_0(B) \subseteq L_1(B) \subseteq \cdots,$$

such that when Alf has a position in $W_k(A)$, he can win in k moves or less. Thus, when Alf has an original position lying in some $W_k(A)$, he can always win; if this is not the case, Betty must always struggle to prevent Alf from achieving a position in such a set. The winning positions for Betty can be found in the same way, and in the remaining ones a draw can always be achieved.

In principle this whole discussion is perfectly fine, and it shows how a little graph theory may be used to analyze all game positions. If it were always practicable, all these games between two persons would be reduced to a triviality, a sort of tic-tac-toe where we should always know what is a bad move and what is a good one. Fortunately, there is no likelihood that our beloved game of chess will suffer this dismal fate, although computers have been developed which will play it to quite a high standard.

We conclude with a couple of simple examples. In the first game there is a pile of matches on the table. The players Alf and Betty move in turn by picking a number of matches, say from 1 to 5, from the heap. The winner is the player who takes the last match. Alf can reach the winning position 0 when there are $W_1(A) = 1, 2, 3, 4, 5$ matches left. Only in one position—when there are six matches—is Betty compelled to play into $W_1(A)$. In turn, this number 6 can be reached by Alf when there are $7, 8, 9, 10, 11$ matches left. Therefore, the set $W_2(A)$ consists of all numbers from 1 to 11 with 6 excluded, while $L_2(B) = 0, 6, 12$. By continuing this way, we conclude that Alf can win from all numbers not divisible by 6. In the first move he leaves a number of matches divisible by 6, and in each subsequent move he takes a number of matches such that they, together with those just taken by Betty add up to 6. If the number of matches in the original pile is divisible by 6, and if Alf makes the first move, Betty can always win in a similar fashion.

Our second example deals with the game of Nim. In its simplest form, there are three piles of playing sticks on the table. In each move, a player selects a pile and takes at least one stick from it; he may take them all. Again, the player who takes the last stick is the winner. In this case we shall be content to describe only the winning positions. This is an exercise in representing numbers in the binary system. You are probably already familiar with this expansion of integers according to the powers of the number 2; the first few representations are

$$
\begin{aligned}
1 &= (1) \\
2 &= (1,0) &&= (1 \times 2) + 0 \\
3 &= (1,1) &&= (1 \times 2) + 1 \\
4 &= (1,0,0) &&= (1 \times 2^2) + (0 \times 2) + 0 \\
5 &= (1,0,1) &&= (1 \times 2^2) + (0 \times 2) + 1 \\
6 &= (1,1,0) &&= (1 \times 2^2) + (1 \times 2) + 0 \\
7 &= (1,1,1) &&= (1 \times 2^2) + (1 \times 2) + 1 \\
8 &= (1,0,0,0) &&= (1 \times 2^3) + (0 \times 2^2) + (0 \times 2) + 0 \\
9 &= (1,0,0,1) &&= (1 \times 2^3) + (0 \times 2^2) + (0 \times 2) + 1 \\
10 &= (1,0,1,0) &&= (1 \times 2^3) + (0 \times 2^2) + (1 \times 2) + 0.
\end{aligned}
$$

Less familiar is the *digital addition* which we can perform on these representations. To see how it works, we consider the following two examples:

$$
\begin{array}{cccc}
1 & 0 & 1 & 1 \\
 & 1 & 0 & 1 \\
 & 1 & 1 & 0 \\
1 & 1 & 0 & 0 \\
\hline
0 & 1 & 0 & 0
\end{array}
\qquad\qquad
\begin{array}{cccc}
1 & 1 & 1 & 1 \\
 & 1 & 1 & 0 \\
 & & 1 & 1 \\
\hline
1 & 0 & 1 & 0
\end{array}
$$

We add column-wise as in ordinary addition, but in the sum we write a 0 in a given column if there are an even number of ones, and a 1 if there are an odd number. As you see, this process is different from ordinary addition in the binary system.

Now back to our game of Nim. For any position we find the digital sum of the three numbers in the heaps. When the distributions are 13, 12, 7 or 14, 11, 5, the respective sums are

$$
\begin{array}{r|cccc}
13: & 1 & 1 & 0 & 1 \\
12: & 1 & 1 & 0 & 0 \\
7: & & 1 & 1 & 1 \\
\hline
 & 0 & 1 & 1 & 0
\end{array}
\qquad\qquad
\begin{array}{r|cccc}
14: & 1 & 1 & 1 & 0 \\
11: & 1 & 0 & 1 & 1 \\
5: & & 1 & 0 & 1 \\
\hline
 & 0 & 0 & 0 & 0
\end{array}
$$

We call a position a *zero-position* if the digital sum contains only zeros, as in the second example; otherwise we have a *regular position*. The winning positions for Alf are the regular positions; he loses in the zero-positions.

The proof is simple. When Alf is in a regular position, he can take out a number of sticks such that he leaves a zero-position for Betty. This he can do by selecting a heap whose number has a 1 in the column of the first 1 in the sum; from this heap he removes a number of sticks so calculated that the new digital sum is zero. For instance, in our first example he may select the heap with $13 = (1,1,0,1)$ sticks, and change it into $11 = (1,0,1,1)$ sticks. In this particular case, he could also have taken sticks from either one of the two other heaps; changing 12 to $10 = (1,0,1,0)$, or 7 to $1 = (0,0,0,1)$, would have accomplished his aim just as well.

When Alf has placed Betty into a zero-position, anything that Betty does will bring Alf back into a regular position; for, any number of sticks that Betty removes will change at least one of the digits in one of the numbers, and hence in the digital sum. The game will end by Alf bringing Betty into the zero-position where all piles are empty. Thus, the situation is somewhat similar to that in the first example.

It is clear that the same method can be applied when there are more than three heaps. Let us observe also that in discussing Nim we did not start with the empty position and work backward to find the winning positions. This could have been done, and with the same result, of course. However, it was easier to make use of the fact that the solution was known, and just to verify that it was correct.

Problem Set 6.2

1. Play a game of Nim with one of your friends (or enemies).
2. Determine the first winning and losing sets $W_1(A)$, $W_2(A)$, $L_1(B)$, $L_2(B)$, in Nim.
3. Examine the following game: there are 2 heaps of matches, and each player has the choice of taking a match from one or the other, or one from each. The winner takes the last match or matches. Find the sets $W_1(A)$, $W_2(A)$, $L_1(B)$, $L_2(B)$. Can you find a general rule for the winning positions?

6.3 The Sportswriter's Paradox

After the football season is over, and after the incidents of the games and the peculiarities of teams and players have been hashed and rehashed until the subject is exhausted, some sportswriter almost always brings up the startling fact that the strongest team was not the supermen of Superb College (S), but rather the dismal group from Rien du Tout (R). This he demonstrates by exhibiting a sequence of games in which R actually won over A, then A over B, and so on, until S is reached at the end of the line.

To analyze when such directed arcs can be found, let us consider the case where there are n teams or contestants

$$a_1, a_2, \ldots, a_n,$$

who have all played against each other. To simplify, we assume that there are no draws. The graph picture is therefore a complete graph (Section 1.2) in which each pair of n vertices is connected by a directed edge. Our first result is the following:

THEOREM 6.1. *In a complete directed graph, there is always a directed path passing through all vertices.*

PROOF. It is sufficient to show that if we have a directed path

$$P: \quad a_1a_2, a_2a_3, \ldots, a_{k-1}a_k,$$

passing through some of the vertices, then we can always find a directed path which passes through one more vertex. So let us add an arbitrary vertex a_{k+1} to those already in P, and consider the various possible directions of the edges connecting a_{k+1} to the vertices in P.

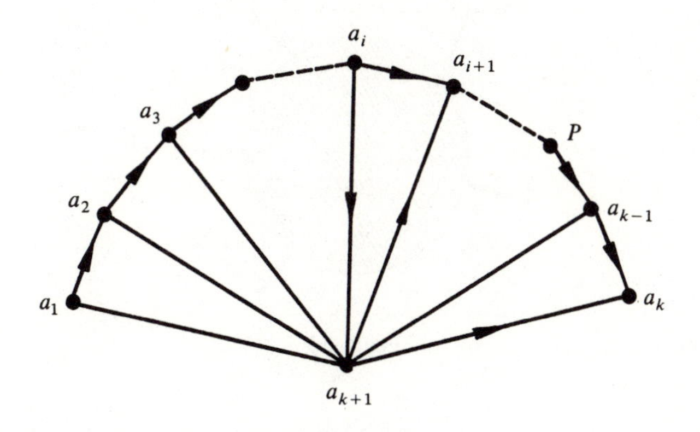

Figure 6.4

If there is an edge directed from a_k to a_{k+1}, we can continue P to a_{k+1}. We suppose therefore that the edge $a_{k+1}a_k$ is directed from a_{k+1} (see Figure 6.4), and we look, consecutively, at the edges connecting a_{k+1} to $a_{k-1}, a_{k-2}, \ldots, a_1$. Suppose that at least one of these edges is directed toward a_{k+1} and let $a_i a_{k+1}$ be the first such edge we come to in the sequence $a_{k+1}a_{k-1}, a_{k+1}a_{k-2}, \ldots, a_{k+1}a_{i+1}, a_i a_{k+1}$. Then, clearly there will be a pair of consecutive edges

$$a_i a_{k+1}, \quad a_{k+1}a_{i+1},$$

the first directed from a_i to a_{k+1}, and the second directed from a_{k+1} to a_{i+1}. In this case, we have the directed path

$$a_1, \ldots, a_i, a_{k+1}, a_{i+1}, \ldots, a_k.$$

There remains the case where all edges are directed from a_{k+1} to P. Then we can begin a path through the edge $a_{k+1}a_1$, and continue through P. This is the only instance where it is necessary to change the initial vertex of the path. □

This result shows that after the games are all completed, we can always arrange the contestants in a directed victory path. The theorem, however, does not quite solve what we had in mind with regard to the sportswriter's paradox, where it was required to take some *fixed* vertex a_1 and draw a directed path from it through all the other vertices. This does not follow from the preceding argument, since in enlarging the arc P we might be compelled to change its initial vertex.

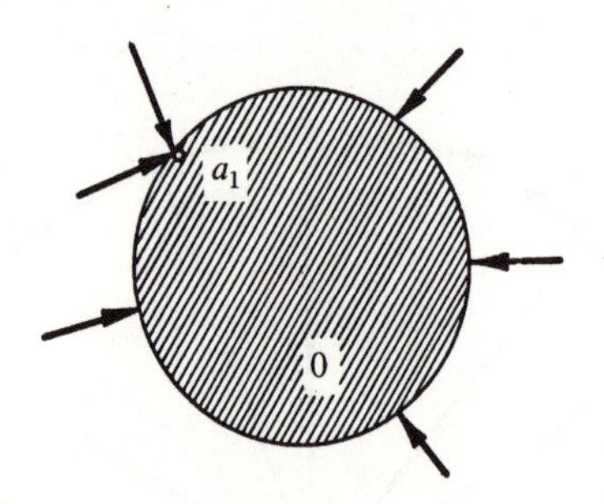

Figure 6.5

In fact, we cannot always find such an arc. This is evident when a_1 is an *outclassed individual*—that is, one who has had nothing but losses in the tournament. More generally, such an arc from a_1 cannot be found if a_1 belongs to an *outclassed group O* whose members have competed only among themselves because no member was able to achieve a victory over an outside opponent (Figure 6.5). In the graph representation, this means that there are only directed edges pointing towards O and none pointing away from O. But we can show the following result:

THEOREM 6.2. *If a_1 does not belong to an outclassed group, there is always a directed path from a_1 through all vertices.*

PROOF. We construct a path P from a_1, making it as long as possible. The proof of Theorem 6.1 shows that, if there is any vertex a_{k+1} to which there is directed an incoming edge from P, then P can be enlarged without changing the initial vertex a_1. Thus, when P cannot be enlarged, its vertices form an outclassed group; but since a_1 does not belong to an outclassed group, P must run through all vertices. □

Our theorem shows that, when there are no outclassed groups, there are directed paths through all vertices from any initial vertex. More specific is the next result:

THEOREM 6.3. *If a directed complete graph has no outclassed groups, then there is a directed cycle through all vertices.*

PROOF. If P is a path through all vertices to the endpoint a_n, then there must also be outgoing edges from a_n, since it cannot be an outclassed vertex. Such an edge $a_n a_i$ must go to a previous vertex a_i on P, so that there must be some directed cycle (Figure 6.6).

Let

$$C = a_1 a_2, \ldots, a_{k-1} a_k, a_k a_1$$

be such a directed cycle which does not include all of the vertices. Suppose that there is some vertex a_{k+1} such that there are edges between C and a_{k+1} in both directions. Then, as in the proof of Theorem 6.1, there are adjacent edges to C in both directions, and we can enlarge C by replacing the edge $a_i a_{i+1}$ by $a_i a_{k+1}$ and $a_{k+1} a_{i+1}$ (see Figure 6.7).

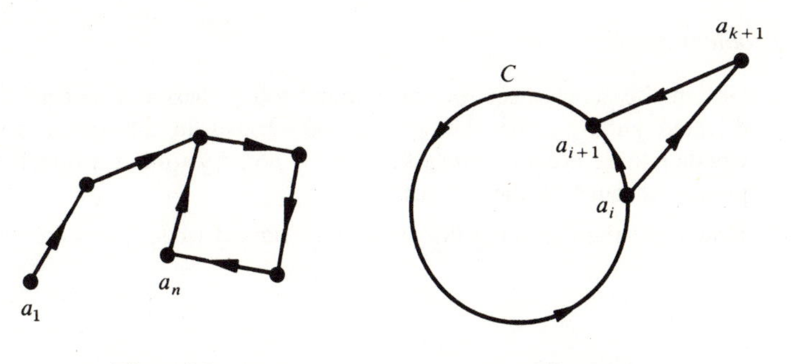

Figure 6.6 Figure 6.7

This leaves us with the case where the vertices outside C fall into two types: *m-vertices* from which all edges are directed to C, and *n-vertices* from which all edges are directed from C. There must be vertices of both these types, for if there were no m-vertices then the set of all n-vertices would form an outclassed group, while if there were no n-vertices, then the vertices in C would be such a group (see Figure 6.8).

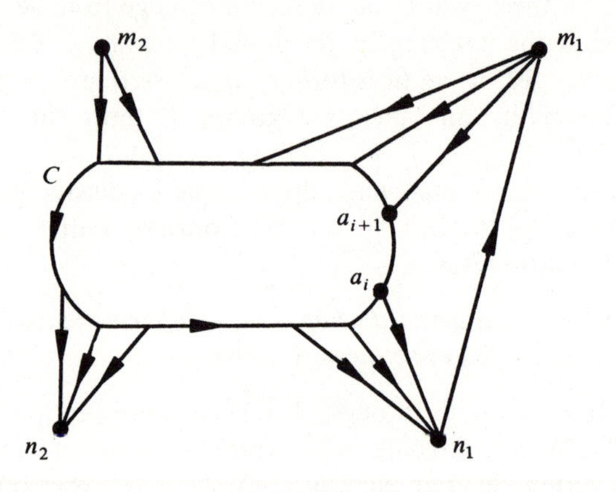

Figure 6.8

The m-vertices must be connected by edges to the n-vertices. There must also be at least one edge n_1m_1 from an n-vertex to an m-vertex, for otherwise the set of n-vertices would be outclassed. But then we can again enlarge C by replacing the edge a_ia_{i+1} by the three edges

$$a_in_1,\ n_1m_1,\ m_1a_{i+1}.$$

In this manner, we can steadily enlarge C until it includes all vertices.

□

Problem Set 6.3

1. Obtain the score sheet for some round-robin chess tournament. Find a directed path passing through all the vertices in the graph. Examine whether there are any outclassed sets; if not, try to find a directed cycle passing through all the vertices.
2. How must the preceding discussion be modified when draws are allowed?

CHAPTER SEVEN

Relations

7.1 Relations and Graphs

So far we have discussed a variety of uses of graphs. Applications to everyday problems and to games and puzzles were considered. Our choice of topics had the advantage that we could deal with well-known and simple concepts. In this chapter, we shall strive to make clear that graphs are closely related to (indeed, are only a different way of formulating) some of the most fundamental concepts of mathematics in general.

A mathematical system, as we usually encounter it, consists of a set of objects or elements. For instance, we deal commonly with numbers and these may belong to more or less general types; we may discuss the set of integers, the positive numbers, the rational numbers, real numbers, imaginary numbers, or complex numbers. In algebra, we are concerned with elements which can be added, subtracted, multiplied, and so on. In geometry, we ordinarily have before us a set of points or special categories of points like straight lines, circles, planes, etc. In logic, we deal with the properties of statements of various kinds.

To construct a mathematical theory we need more than these elements; we need *relations* between them. Let us illustrate this: in the case of numbers, we have equal numbers a and b; in formal mathematical terminology, we write $a = b$. We also have numbers a and b which are different, and we write $a \neq b$. The symbol $a > b$ denotes that a is greater than b, and $a \geqslant b$ means that a is either greater than b, or equal to b. When we deal with integers and a divides b, we write $a|b$.

In geometry, two objects (say, triangles) A and B may be congruent, in which case we write $A \cong B$; or one may include the other, and we write $A \supset B$. Two straight lines may be parallel, $A \| B$, or they may be orthogonal (intersecting in a right angle), $A \perp B$. In logic, one statement or assertion may imply another, $A \rightarrow B$. In set theory, the relation $a \in S$ between an element a and a set S means that a is an element of the set S.

All of these relations concern two objects, and are therefore called *binary relations*, or simply *relations* (for short). There are other relations; for instance, a *ternary relation* concerns three objects. As an example of such a relation we may take: A lies between B and C.

The importance of relations in mathematics makes it necessary to arrive at a general definition for them. For a general (binary) relation with the symbol R we write

$$aRb,$$

and say that *b is in the relation R to a*. This means that b belongs to some special set R_a which is determined for each a by the relation. So, for instance, the relation $a > b$ means that b belongs to the set of all numbers which are less than a. For integers a and b, the relation $a|b$ (or a divides b) means that b belongs to the set of all integers which are multiples of a. Therefore, in general, aRb is another way of expressing that b belongs to the set R_a which R associates with a.

Let us return to our graphs. Actually, each directed graph G defines a relation on its vertex set. This relation we may write

$$aGb,$$

and it signifies that there is a directed edge from a to b in G. The corresponding set $R_a = G_a$, associated with a under the relation, is then the set of all those vertices b in G to which there is an edge from a. Thus, to state that there is an edge ab in G is the same as saying that the relation aGb holds.

As a consequence, we might be inclined to consider the theory of graphs as a special part of the theory of relations. Actually the two theories are co-extensive. Suppose that a relation R is defined in a set S, so that there is a set R_a associated with each a in S. Then we can construct a graph G for R simply by drawing an edge from a to each vertex in R_a (see Figure 7.1).

Since the theory of relations and the theory of graphs are only different aspects of the same concepts, we may ask why mathemati-

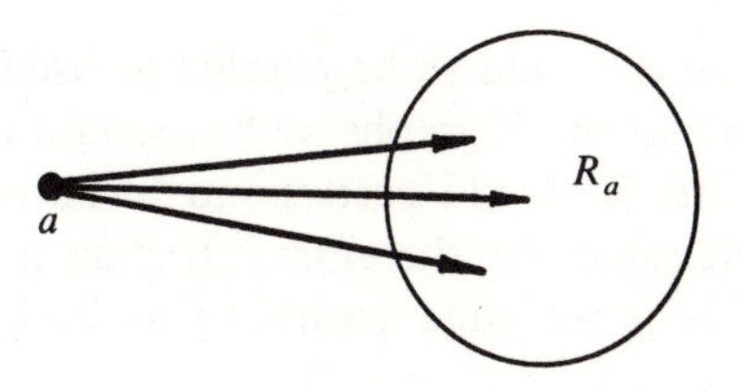

Figure 7.1

cians maintain a distinction between them. This is partly due to habit and tradition, just as in analytic geometry one talks about a straight line while in algebra the same concept is a linear equation in two unknowns. But there is actually a difference in the methods of the two disciplines, although not a sharp one, of course. The reason is that in relations we deal mostly with infinite sets R_a; as an example, take the relation $a > b$ for all real numbers. This means that in the corresponding graph picture there should be an infinite number of vertices and an infinite number of edges from each vertex. It is extremely difficult to have any intuitive perception of the properties of such a graph. When we discussed graph problems in the preceding chapters, it was of great help to be able to reason geometrically about the vertices and edges connecting them. For many types of relations, these arguments lose their lucid character; indeed, they may not be valid for infinitely many elements, and so one is forced to introduce other types of proofs for relations in infinite sets.

Problem Set 7.1

1. List some relations other than those mentioned. Try to make up some of your own.
2. For the set of numbers $\{2,3,4,5,6\}$, draw the graph and list the sets R_x for each of the relations:

$$(a)\ x > y; \qquad (b)\ x \neq y; \qquad (c)\ x|y.$$

7.2 Special Conditions

New points of view usually produce new observations. There are certain aspects of relation theory which should also be introduced in graph theory to make the parallelism more complete.

For any relation R, it may happen that an element is in this relation to itself:

$$aRa.$$

For instance, a line A is said to be parallel to itself, $A \| A$; a number satisfies $a \geqslant a$, and so on. In graphs we have so far made no provision for this special case. It should correspond to an edge (a, a) having both endpoints the same. We therefore introduce a *loop* (a, a) in the graph picture; this is an edge returning to itself at the vertex a (Figure 7.2).

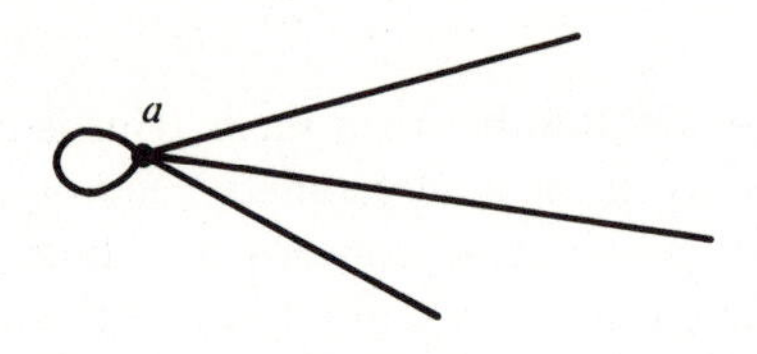

Figure 7.2

A relation R such that aRa holds for every a is called a *reflexive relation*. In the graph picture, this means that there is a loop at every vertex. As examples, we have as above the relation for lines, $A \| B$; or for numbers, $a \geqslant b$.

If the relation aRa does not hold for any element, we say that R is an *anti-reflexive relation*. This is equivalent to the property that the corresponding graph has no vertices with loops. As an example, we take the relation for orthogonal lines, $A \perp B$—that is, A and B intersect at an angle of 90°.

For any relation R, we can define the *converse relation* R^* by writing bR^*a whenever aRb holds. So, for instance, the relation $a|b$ (that is, a divides b) has for its converse relation $b|^*a$ (or b is a multiple of a). Frequently there is a special symbol for R^*: the relation a is greater than b $(a > b)$ has the converse $b < a$ (or b is less than a). The relation $a \in A$ (a is an element of A) has for its converse $A \ni a$ (or A contains a).

The definition of the converse relation shows that if there is an edge (a, b) in the graph G corresponding to R, then there is an edge (b, a) in the graph G^* corresponding to R^*. In other words, G^* is the *converse graph* to G—that is, the same graph but with oppositely directed edges.

It may happen that for a relation R one has simultaneously

$$aRb \quad \text{and} \quad bRa,$$

for a pair of elements a and b. In the graph picture, we should then have two edges ab and ba, one in each direction. But in graphs, we can replace such a pair by a single edge without direction, just as in the case of two-way streets.

Some relations have the property that one of the relations aRb and bRa always implies the other; such relations are called *symmetric*. As examples let us mention the parallel relation, $A \| B$; the orthogonal relation, $A \perp B$; and the equality relation, $A = B$. According to the remarks we have just made, we can say:

a symmetric relation has a graph with undirected edges; conversely, a graph with undirected edges defines a symmetric relation.

There are some relations R in which one of the relations aRb and bRa can never hold when the other is satisfied. As an example, take $a > b$. Such relations are called *anti-symmetric* or *asymmetric*. Their graphs have no undirected or two-way edges between any two vertices; furthermore, there are no loops—that is, these relations are antireflexive.

There is another property which plays an important role in relation theory. We say that a relation R is *transitive* if the two conditions aRb and bRc imply aRc. Among the examples, let us list the relations, A is parallel to B, $A \| B$; a is equal to b, $a = b$; a is greater than b, $a > b$; a divides b, $a|b$. On the other hand, the relations $A \perp B$ and $a \neq b$ are not transitive.

A transitive relation has the following characteristic graph property: for any pair of edges $(a, b), (b, c)$ there exists a *resultant* edge (a, c) (see Figure 7.3).

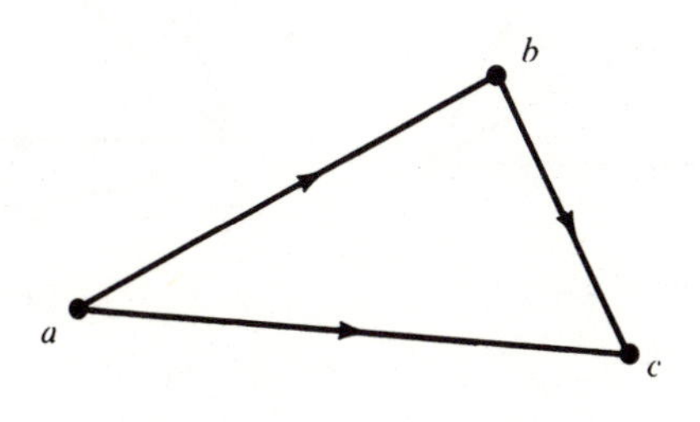

Figure 7.3

By repeated use of this property, we conclude that when there is a directed path from a vertex x to another y, then there is also an edge xy.

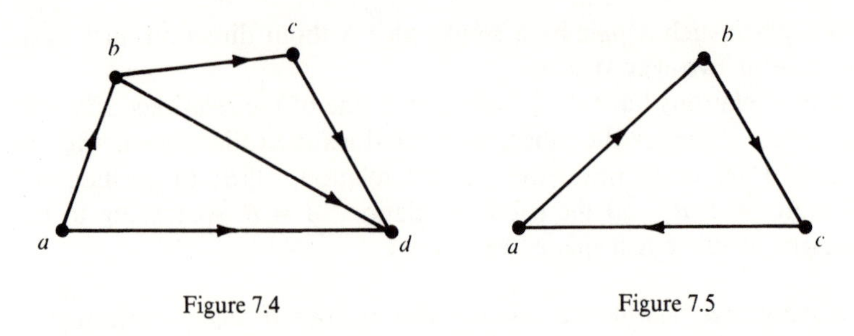

Figure 7.4

Figure 7.5

Suppose finally that we have a graph G with directed edges which is not transitive. For example, the graph in Figure 7.4 has no edge between the vertices a and c. The graph in Figure 7.5 has an edge ca, while the resultant edge of ab and bc would be the edge ac. In any case, we can always make a directed graph G transitive by adding directed edges until, for each pair of consecutive edges, the resultant edge has been included. The new graph G^T so obtained is called the *transitive closure* of G. Figures 7.6 and 7.7 show the transitive closures of the graphs in Figures 7.4 and 7.5, respectively. Observe that when the resultants ac of ab and bc, cb of ca and ab, and ba of bc and ca, are added to the graph in Figure 7.5, then the transitive closure so obtained is the graph of a symmetric relation.

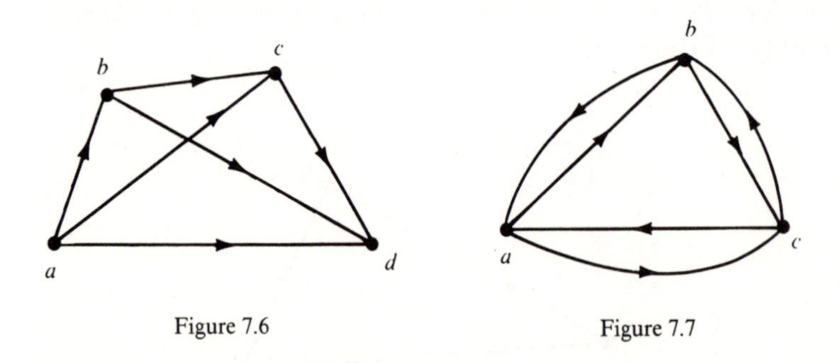

Figure 7.6

Figure 7.7

Problem Set 7.2

1. Examine some other relations with respect to the properties mentioned above.
2. Examine the family relations:
   (a) A is a descendant of B; (b) A and B have a common forefather.
3. Form the transitive closures of the graphs in Figures 5.1 and 5.2.

7.3 Equivalence Relations

Among the many types of mathematical relations, the equivalence relations play an outstanding role. An *equivalence relation*, denoted by the symbol $\sim$, is defined by three properties:

reflexivity: $a \sim a$.
symmetry: if $a \sim b$, then $b \sim a$.
transitivity: if $a \sim b$ and $b \sim c$, then $a \sim c$.

An example which comes to mind immediately is equality, $a = b$. Indeed, the equivalence relations have properties similar to equality, and in many instances one considers an equivalence as a new kind of equality; a good example is geometric congruence, $A \cong B$, for two figures.

In number theory we use another type of equivalence relation which is also called a *congruence*. When a, b and m are three integers, we write

$$a \equiv b \pmod{m}$$

whenever the difference $a - b$ is divisible by m; in other words, when

$$a = b + km,$$

where k is some integer.

In number theory, we express this relation in words as follows: the number a is congruent to b modulo m. As examples, let us take

$$11 \equiv 2 \pmod{3} \quad \text{and} \quad -7 \equiv 19 \pmod{13}.$$

For systematic reasons, we also use the notation $a \equiv b \pmod{m}$ in cases where we commonly use other terms. For instance, the congruences

$$b \equiv 0 \pmod{2}, \quad c \equiv 1 \pmod{2}$$

mean, respectively, that b is an even number and that c is an odd number. Also,

$$a \equiv 0 \pmod{m}$$

means that a is divisible by m.

To prove that the congruence $a \equiv b \pmod{m}$ satisfies the three conditions for an equivalence relation is quite simple:

reflexive: $a \equiv a \pmod{m}$, since $a = a + 0m$;
symmetric: if $a \equiv b \pmod{m}$, then we have $a = b + km$, and therefore $b = a + (-k)m$, so that $b \equiv a \pmod{m}$;

transitive: if $a \equiv b \pmod{m}$ and $b \equiv c \pmod{m}$, then

$$a = b + km,\ b = c + lm, \text{ for some integers } k \text{ and } l,$$

and therefore

$$a = c + lm + km = c + (l + k)m,$$

so that $a \equiv c \pmod{m}$.

For $m = 0$, the congruence $a \equiv b \pmod{m}$ reduces to ordinary equality, $a = b$.

An equivalence relation can also be interpreted in a different manner, as we shall now see. First we notice that an equivalence relation is always defined for the elements of some set S; two elements are either equivalent or non-equivalent. For the congruence relation $a \equiv b \pmod{m}$ we deal with the set S of all integers, whereas the parallel relation $A \| B$ is an equivalence relation defined for the set of all straight lines in the plane or in space.

Suppose that an equivalence relation $a \sim b$ is defined for the elements in a set S, and consider all elements b equivalent to a. These elements form a set B_a which is part of S. The sets B_a correspond to the relation sets R_a we introduced earlier for general relations, but in this case we prefer to call them the *equivalence classes* of our equivalence relation.

What are the properties of these classes? Since the relation is reflexive, $a \sim a$, the equivalence class B_a of a contains a. Suppose now that b is an element in B_a, so that $a \sim b$, and let c be an element in the equivalence class B_b of b, so that $b \sim c$ (see Figure 7.8). Since the relation is transitive, we have $a \sim c$, so c belongs to B_a. This means that the whole equivalence class B_b is contained in B_a. But if $a \sim b$, then $b \sim a$, since the relation is symmetric; we conclude in the

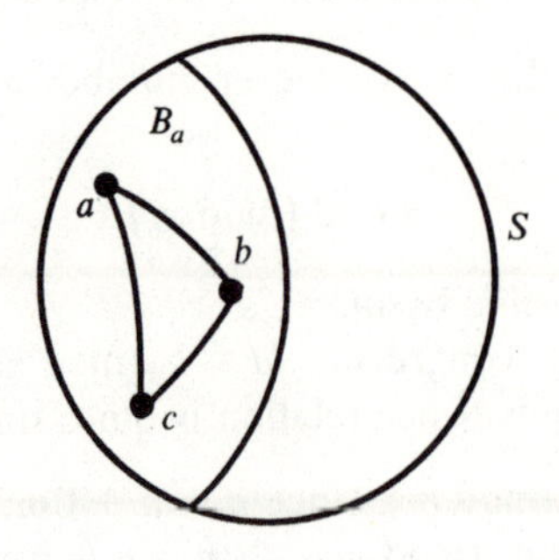

Figure 7.8

same manner that B_a is contained in B_b. This shows that $B_a = B_b$ if $a \sim b$.

Next take two elements a and b in S that are not equivalent. Then the two classes B_a and B_b are *disjoint*—that is, they have no elements in common. For, if an element c belonged to both B_a and B_b, it would follow that

$$a \sim c \quad \text{and} \quad b \sim c,$$

and so $a \sim b$, contrary to assumption.

Since each element belongs to just one of the classes, we have decomposed the whole set S into disjoint equivalence classes (see Figure 7.9). Each class consists of a set of equivalent elements.

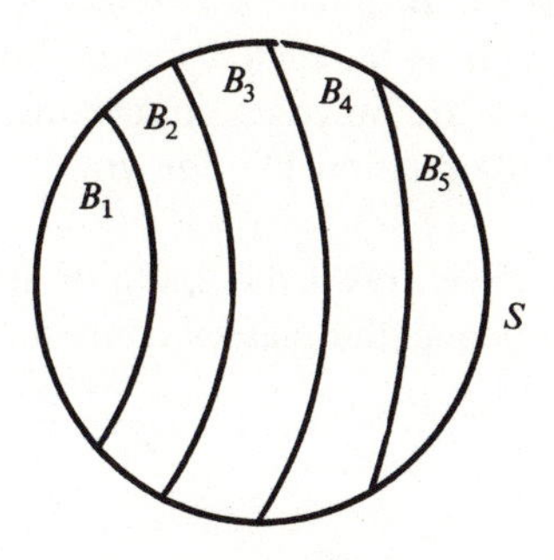

Figure 7.9

As an example, let us take first the family S of all straight lines in the plane and the relation $A \| B$; here each equivalence class consists of all lines having the same *direction*.

Secondly, the number theory congruence divides all integers into equivalence classes; these are called *residue classes* (mod m). Each class consists, respectively, of the numbers

$$B_0 = km, \quad B_1 = 1 + km, \quad \ldots, \quad B_{m-1} = (m-1) + km.$$

In other words, B_r consists of the numbers which leave the remainder r when divided by m. For $m = 2$, we find the division of the integers into even and odd numbers; when $m = 3$ we have the three different types of numbers

$$3k, \qquad 1 + 3k, \qquad 2 + 3k.$$

We saw that an equivalence relation defines a decomposition of all elements in the set S into disjoint classes, as in Figure 7.9. But suppose, conversely, that we have some such decomposition of S into sets B_i without common elements; this is usually called a *partition* of the set S. Then we can define an equivalence relation in S, simply by putting $a \sim b$ whenever a and b belong to the same set B. It is evident that the three conditions for an equivalence are satisfied. This shows that an equivalence relation, and a partition or decomposition of the set S into disjoint parts, are in reality two aspects of the same thing. Each equivalence gives a disjoint decomposition, and each such decomposition defines a unique equivalence.

Let us take a non-mathematical illustration. Two individuals have the same nationality if they are citizens of the same state; conversely, the nationalities divide humans into classes.

Let us not forget a third aspect of equivalence relations—their graphs. When S is taken to be a set of vertices, all those vertices which belong to a block B_i must be connected by edges, so that we obtain a complete graph G_i for each B_i; in addition, there is a loop at each vertex. Since there are no edges connecting two different classes, the graph of the whole relation has the graphs G_i for its connected components.

In Figure 7.10 we have drawn the graph of an equivalence relation in a set with 12 elements; the classes contain, respectively, 5, 4, 2, 1 elements.

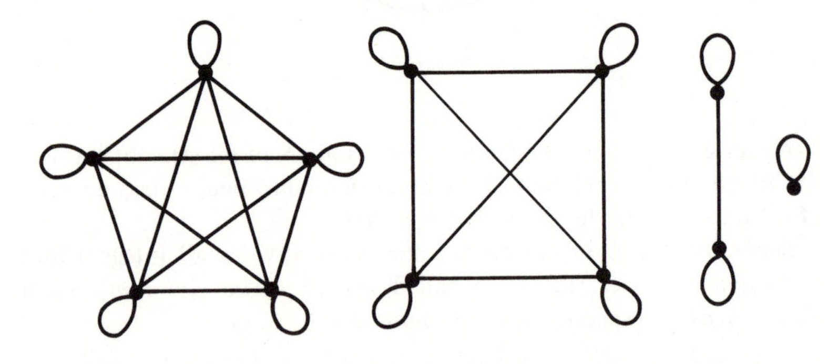

Figure 7.10

Problem Set 7.3

1. Find other examples of equivalence relations.
2. Prove that the congruences $a \equiv b \pmod{m}$ can be added, subtracted and multiplied; that is, if

$$a \equiv b \pmod{m} \quad \text{and} \quad c \equiv d \pmod{m},$$

then

$$a \pm c \equiv b \pm d \pmod{m} \quad \text{and} \quad ac \equiv bd \pmod{m}.$$

3. Two real numbers are called *associated* when they have the same absolute value, $|a| = |b|$. Show that this is an equivalence relation, and determine the elements in each equivalence class.

7.4 Partial Orders

To illustrate another type of basic mathematical relation, we suppose that we have a set S of elements of some kind; these elements may be numbers, points, or even the human beings in the world. Within this set S, we consider some family of smaller sets or subsets $A, B, \ldots$. If, for instance, S consists of all the real numbers, each subset in our family may be the numbers in a certain interval; if S is the set of points in the plane, the sets $A, B, \ldots$ may be all the points on given lines or figures.

For such subsets we have an inclusion relation $A \supseteq B$, which may or may not be fulfilled for a pair of our sets; it signifies that all elements in B also belong to A. The two sets may have the same elements, so that $A = B$. One sees immediately that the following three conditions are fulfilled for an *inclusion relation*:

reflexivity: $A \supseteq A$;
transitivity: if $A \supseteq B$ and $B \supseteq C$, then $A \supseteq C$;
identification: if $A \supseteq B$ and $B \supseteq A$, then $A = B$.

Instead of using the relation $A \supseteq B$, we often apply the *strict inclusion relation*, denoted by $A \supset B$. This expresses as before that B is a subset of A, but the possibility $A = B$ is excluded, so that B is a *proper subset* of A. In this case we have the properties for a *strict inclusion relation*:

anti-reflexivity: $A \supset A$ cannot occur;
transitivity: if $A \supset B$ and $B \supset C$, then $A \supset C$.

A binary relation $a \geqslant b$ which satisfies the conditions for an inclusion relation is called a *partial order*. It satisfies therefore the axioms:

$a \geqslant a$;
if $a \geqslant b$ and $b \geqslant c$, then $a \geqslant c$;
if $a \geqslant b$ and $b \geqslant a$, then $a = b$.

In the various branches of mathematics, we use inclusion symbols both in the pointed form $a \geqslant b$ and in the rounded form $a \supseteq b$; the latter form is often reserved for set inclusion.

In reality, partial order and set inclusion are only different formulations of the same idea. For each element a in a partial order, there is a relation set R_a consisting of all the elements b for which $a \geqslant b$. Since the relation is reflexive, a belongs to R_a. Two different elements a and b correspond to different sets R_a and R_b; for, if $R_a = R_b$, then $a \geqslant b$ and $b \geqslant a$, and so $a = b$. Now, as we shall show, we have $a \geqslant b$ if and only if $R_a \supseteq R_b$. For, if $a \geqslant b$, then for any element c in R_b, $b \geqslant c$, and so $a \geqslant c$; that is, $R_a \supseteq R_b$. On the other hand, if $R_a \supseteq R_b$, then $a \geqslant b$, since b is in R_b.

In the same way, we introduce *strict partial order* as a relation $a > b$ which satisfies the conditions:

$a > a$ is not possible;
if $a > b$ and $b > c$, then $a > c$.

As above, we verify that $a > b$ if and only if $R_a \supset R_b$, for the corresponding relation sets. The symbol $>$ is used here in a more general sense than in the usual "greater than" meaning for numbers.

Next we turn to the graph picture G of a partial order, particularly in the case where there are only a finite number of vertices. It is not necessary to make any particular distinction between a partial order and a strict partial order; the only difference is that the first graph has loops, whereas the second graph does not. Whenever $a > b$, there is a directed edge ab in G. In Figure 7.11, we have drawn such a strict partial order on eight vertices.

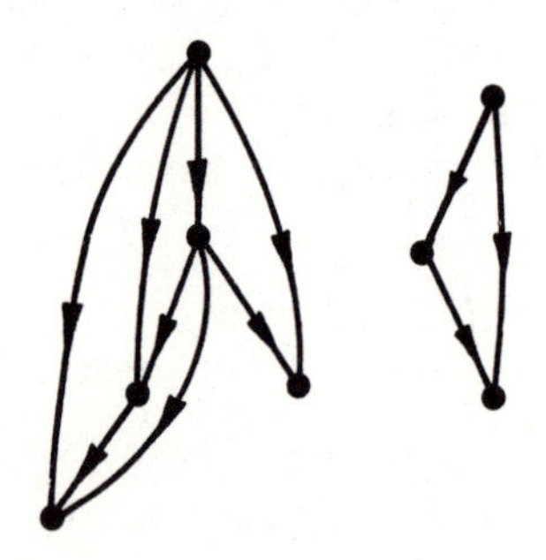

Figure 7.11

We hasten to add that this is not the usual way of presenting a partial order. In the graph above, whenever there are two edges ab and bc there is also an edge ac. We can simplify the diagram by leaving out all such resultant edges. In general, whenever there is a directed path A_{ab} from a to b, there should, strictly speaking, also be an edge ab; but since all such resultant edges are direct consequences of the existence of directed paths, these edges may be considered to be superfluous, and so they can be omitted. When these reductions are made in the graph in Figure 7.11, we are left with the much simpler graph shown in Figure 7.12.

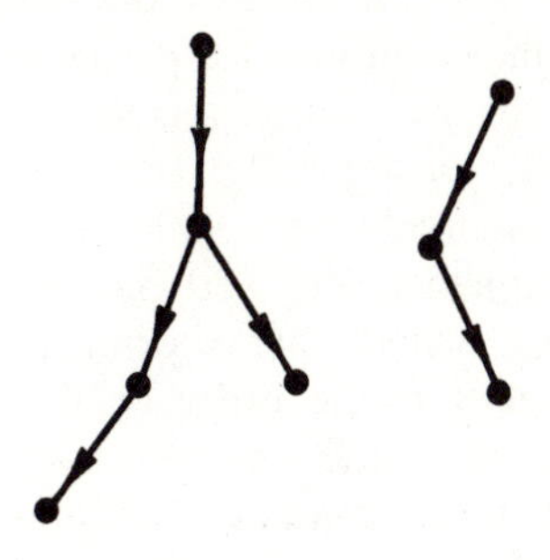

Figure 7.12

The graph of a partial order from which all loops and superfluous edges have been removed is called a *basis graph*. If A_{ab} is a directed path in the basis graph, no directed edge ab can exist, since it would be superfluous. Nor can there be any directed edge ba in the opposite direction, for together with A_{ab} it would produce a directed cycle returning to a, giving $a > a$, contrary to assumption (Figure 7.13). On the other hand, an acyclic directed graph (that is, a directed graph without directed cycles) produces a partial order when we write $a > b$ whenever there is a directed path from a to b.

Figure 7.13

A (strict) *order relation* is a (strict) partial order $a > b$ which, *in addition to the previous conditions*, also fulfills the condition of *com-*

pleteness: for any two different elements a and b, one of the relations $a > b$, $b > a$, is always satisfied. In other words, *all* elements in the set can be 'compared' in an order relation. This is sometimes called a *complete order*, in contradistinction to the partial orders. Analogously, the order relation $a \geqslant b$ is defined by the same completeness condition, in addition to the conditions for partial order.

In an order relation, the relation sets R_a and R_b of two elements must satisfy $R_a \supset R_b$, or $R_b \supset R_a$. In mathematics, as well as in everyday life, we commonly deal with ordered sets; for example, names are ordered alphabetically, and children may be rated according to height, scores, school marks or many other criteria. Most familiar in mathematics is the order of the numbers on the real number line. Here, the relation set R_a consists of all numbers b satisfying $a > b$; that is, R_a consists of all numbers to the left of a on the axis (Figure 7.14).

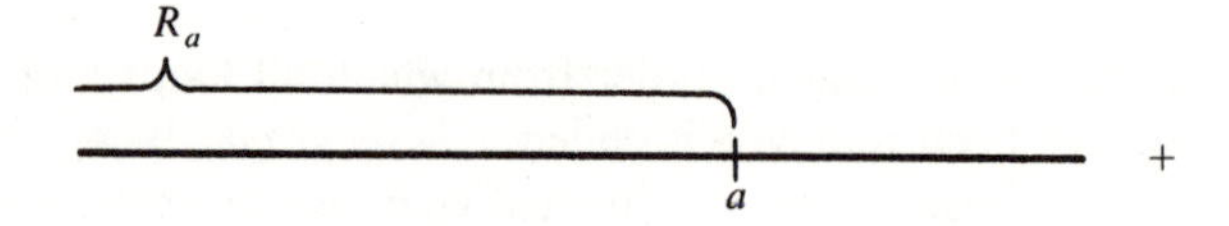

Figure 7.14

Very often, we deal with ordered sets of a *finite* number of elements. Such a set contains an element a_1 smaller than all the others, a next element a_2 greater than a_1 but smaller than the others, and so on up to the greatest element a_n. This means we can order the elements, just as we order the integers $1, 2, \ldots, n$ according to size. In more formidable mathematical terms, this would be expressed: any ordered set with n elements is order isomorphic to the integers $1, 2, \ldots, n$.

Problem Set 7.4

1. Draw the complete graph and the basis graph for the order of the four numbers 1, 2, 3, 4.
2. What is the basis graph for the order of the n numbers $1, 2, \ldots, n$?
3. Prove that the division relation $a|b$ for integers is a partial order. What is the distinction in this case between the partial order and the strict partial order?
4. Take all subsets of a set S with the three elements a, b, c. How many subsets are there? Construct the graph and the basis graph for the corresponding partial order. Why is it desirable to introduce also an empty subset $\varnothing$ with no elements?

CHAPTER EIGHT

Planar Graphs

8.1 Conditions for Planar Graphs

As we have already explained in Section 1.4, a planar graph is a graph which can be drawn in the plane so that the edges have no intersections except at the vertices. We also gave a number of illustrations of planar graphs. In Section 1.5, we analyzed the problem of the three houses and the three wells, and explained why the corresponding graph could not be planar. The graph of the problem (Figure 1.16) can be drawn in many ways, as is possible for all graphs. When we say "the graph," we mean any graph isomorphic to a particular graph describing the situation. Therefore, the statement "the graph in Figure 1.16 is not planar" means it has no planar isomorph. For instance, the vertices may be placed in a hexagon as in Figure 8.1. The intersection in the center is not a vertex; the edges should be considered to pass over each other at this point.

There is even a graph with only 5 vertices which is not planar—namely, the complete graph on 5 vertices (Figure 8.2). Why this graph is not planar may be made clear by reasoning similar to that used in Section 1.5 to show that the graph in Figure 8.1 (or Figure 1.16) is not planar. The vertices in any representation of the graph must lie on a cycle C in some order—say, *abcdea*. There is an edge *eb*, and in our planar graph we have the choice of putting it on the inside or the outside of C. The argument is similar in both cases;

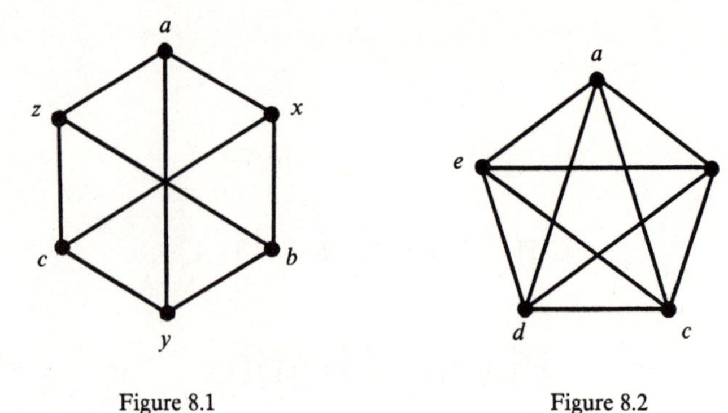

Figure 8.1

Figure 8.2

let us put it on the inside as in Figure 8.3. From *a*, there are edges to *d* and *c*. Since the edge *eb* prevents them from being inside *C*, they must both be on the outside. The edge *db* can be drawn only inside *C*, since the edge *ac* blocks the access to *b* from the outside. But then we have no way of drawing the last edge *ce* in the plane, for it cannot be placed inside *C* due to the edge *db*, nor outside *C* due to the edge *ad*.

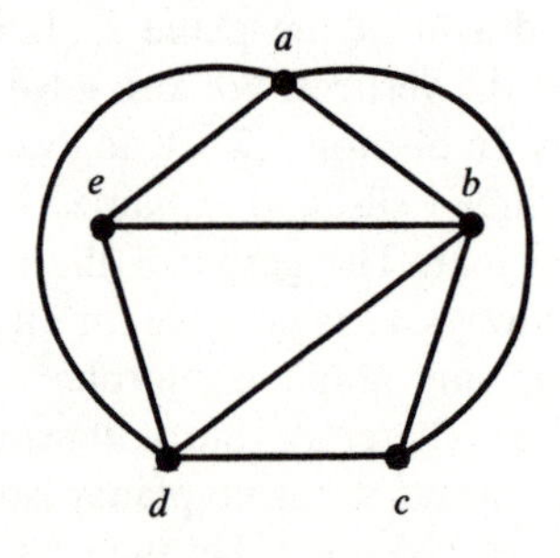

Figure 8.3

We have discussed in some detail the special graphs representing the three houses and the three wells (Figure 8.1) and the complete graph with 5 vertices (Figure 8.2), because they play a particular role in determining whether or not a given graph is planar. To state a criterion for planar graphs due to the Polish mathematician Kuratowski (1930), we must explain what we mean by *expanding* and *contracting* a graph.

Suppose we add new vertices on some edges of a graph, so that these edges become paths consisting of several edges; this operation we call *expanding* the graph. In Figure 8.4 we have illustrated the expansion of a graph (a) with four vertices to a graph (b).

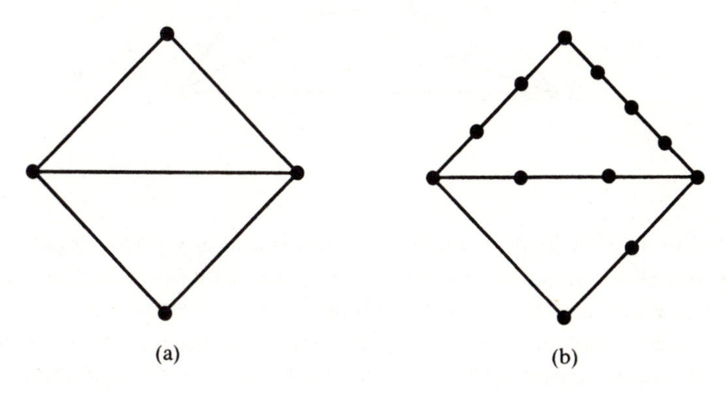

Figure 8.4

Conversely, suppose we have a graph, as in (b), consisting of divided paths with no edges at the intermediate vertices. By the reverse operation, it can be *contracted* to a graph in which the original paths become edges. So in Figure 8.4, the graph (b) can be contracted to the graph (a) by dropping the inner vertices of the paths. We are now able to state the theorem of Kuratowski:

A graph is planar if and only if it does not contain within it any graph which can be contracted to the pentagonal graph (Figure 8.2) *or the hexagonal graph (Figure* 8.1).

The proof is not very difficult, but it is somewhat involved and would require more space than we can afford here.

Let us add a few other observations about planar graphs. We tried in the preceding, for instance in Figure 1.17 or Figure 8.3, to find a planar representation of a graph by drawing its edges as ingeniously bending curves. Actually, this is not essential. It can be shown that any planar graph can be drawn in the plane in such a manner that all edges are straight lines—provided, of course, that no pair of vertices is connected by more than one edge. As an illustration, we may take the graph in Figure 8.3 and represent it as in Figure 8.5.

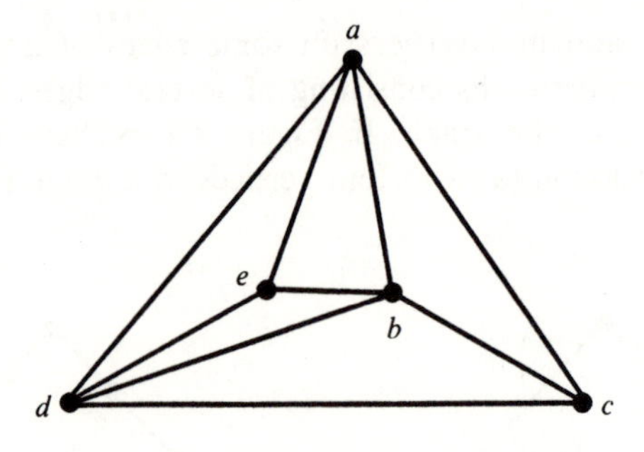

Figure 8.5

Another fact which we should mention is that any planar graph can just as well be drawn with vertices and edges lying on the surface of a sphere. There are many ways of obtaining such a representation. We can, for instance, use a *stereographic projection*; this is a method often introduced by cartographers, particularly for maps of regions of the earth around the poles.

Let P be the plane in which our graph is located. We place a sphere S so that it touches P at its Antarctic Pole A (Figure 8.6).The North Pole N is taken as the projection center. To any point p in the plane P, we draw the line from N; this line intersects S in a point s. To each p we associate the corresponding point s on S. By the reverse procedure, any graph on the sphere S can be projected back upon a plane P tangent to S; the only point on S that has no image on P is the projection center N.

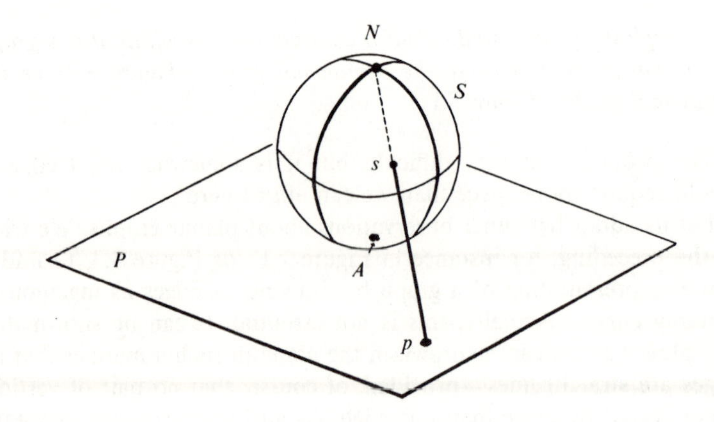

Figure 8.6

Problem Set 8.1

1. Take away the edge ay from the graph in Figure 8.1. Draw the remaining graph in the plane with straight-line, non-intersecting edges.
2. Try to find all graphs with 6 vertices which are not planar.

8.2 Euler's Formula

We shall now examine planar graphs which form a *polygonal net* in the plane. By this, we mean that the edges in the planar graph G form a set of adjoining polygons in the plane, dividing it up into polygonal pieces, as we have indicated in Figure 8.7.

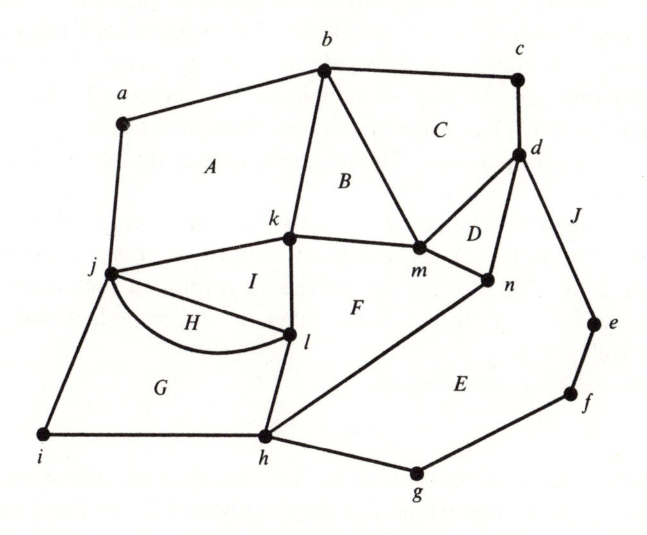

Figure 8.7

To avoid any misunderstanding, let us emphasize that, in contrast to ordinary usage, when we talk about polygons we do not necessarily mean that the edges must be straight lines. They may be any kind of non-self-intersecting continuous curves which divide up the plane. A very good illustration of a *polygonal graph* is a map of the 48 contiguous states of the U.S.A., showing the division into the various states. Indeed, any map of the border lines between various countries may serve. The boundaries between the states are the edges of the graph, and the states or countries themselves are the polygons.

Now let us return to our polygonal graphs. As we have indicated, they are connected; moreover, we require that no polygon shall completely surround another. The boundary edges of a polygon form a cycle, sometimes called a *minimal cycle*. The part of the plane which it surrounds is a *face* of the graph. There is also a *maximal cycle* C_1, which surrounds the whole graph with all of its faces. A most excellent guiding principle in mathematics consists in introducing such conventions that the formulas become as simple as possible. In our case, it turns out to be advantageous to consider the part of the plane lying outside C_1 to be a face of the graph, with C_1 as boundary. This we call the *infinite face*. You can see that, if we project the whole graph on a sphere as indicated in Section 8.1, there is really no distinction between the infinite face and the others.

Let us illustrate our discussion by the graph in Figure 8.7. Here we have a graph with 10 faces lettered A to J. For instance, Face A has a boundary cycle consisting of the edges ab, bk, kj, ja, while Face H is bounded only by the two edges joining the vertices j and l. The maximal cycle C_1 has edges which run through the letters a to j in order, and then back to a. The infinite face J is the set of all points outside C_1.

For polyhedra in space, there exists an interesting relation first observed by Euler; it is commonly known as *Euler's polyhedral formula*. It is valid also for our polygonal graphs. In such a graph G, we denote by n, m and f, the number of vertices, edges and faces, respectively, of G.

Then Euler's formula states:

$$n - m + f = 2.$$

PROOF. The formula is valid in the simplest case where we have only one polygon consisting of k edges (Figure 8.8). In this case,

$$n = m = k, \quad f = 2,$$

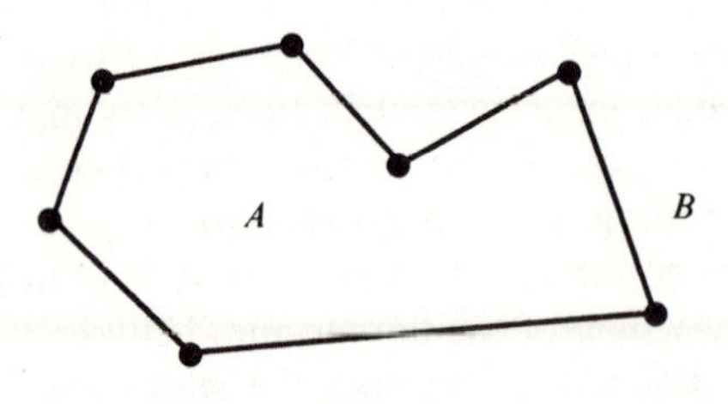

Figure 8.8

and so Euler's formula holds. We shall use mathematical induction to prove the relation in general. We shall now show that, if it holds for graphs with f faces, then it also holds for graphs with $f+1$ faces. Polygonal graphs can be constructed stepwise; in each step, a face is added 'on the outside.' Suppose that G (the solid part of Figure 8.9) is a polygonal graph with n vertices, m edges and f faces, and that the numbers n, m and f satisfy Euler's formula. We add a new face as indicated (see the dashed lines in Figure 8.9) by drawing a path through the infinite face of G, connecting two vertices on the maximal cycle of G. If this path has r edges, we have added $r-1$ new vertices and one new face. But then it is clear that Euler's relation remains valid for the augmented graph, since

$$n - m + f = (n + r - 1) - (m + r) + (f + 1). \quad \square$$

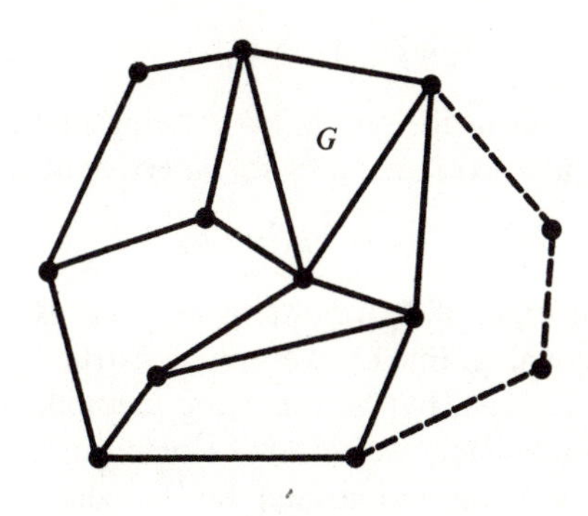

Figure 8.9

Problem Set 8.2

1. Verify Euler's formula for the graph in Figure 1.13.
2. Do the same for the graph formed by the 8×8 squares of a chessboard. Generalize to a board with $k \times k$ squares.

8.3 Graph Relations and Dual Graphs

We shall assume from now on that we deal with polygonal graphs. We write Euler's formula in the form

$$n + f = m + 2.$$

The number of edges of a graph can be obtained by counting the edges at each vertex. Since each edge is counted twice in this manner, we obtain the handshaking lemma,

$$2m = \deg(a_1) + \cdots + \deg(a_n),$$

just as in Section 1.6. For the graph in Figure 8.7, we find $m = 22$.

There is another way of counting the edges of a polygonal graph. For a given graph G, let φ_k denote the number of faces in G bounded by k edges. For example, the graph in Figure 8.7 has

$$\varphi_2 = 1, \quad \varphi_3 = 3, \quad \varphi_4 = 3, \quad \varphi_5 = 1, \quad \varphi_6 = 1,$$

$$\varphi_7 = 0, \quad \varphi_8 = 0, \quad \varphi_9 = 0, \quad \varphi_{10} = 1.$$

In other words, among its 10 faces there is one bounded by two edges, there are three bounded by three edges, and so on.

Since there are no loops in these polygonal nets, there are no faces bounded by only one edge. We therefore have the equation

$$f = \varphi_2 + \varphi_3 + \varphi_4 + \cdots.$$

We count the edges in the graph, by noticing that each edge lies on the boundary of just two faces, and so we arrive at the formula

$$2m = 2\varphi_2 + 3\varphi_3 + 4\varphi_4 + \cdots.$$

In the example in Figure 8.7, we obtain $m = 22$, as before.

For any polygonal graph G, we can construct a new polygonal graph G^*, its *dual graph*, by the following method: within each face, including the infinite face, we select a single point. Two such inner points a and b are then connected by an edge if they belong to neighboring faces with a common boundary edge e, and the new edge from a to b is drawn so that it crosses e but no other edges of the graph. If there are several boundary edges common to the two faces, one new edge is drawn for each. We have illustrated the situation in Figure 8.10, where G consists of solid lines and G^* consists of dashed lines.

The dual graph G^* is of importance in the study of planar graphs; indeed, such dual graphs can be defined only for planar graphs.

The dual graph of a polygonal graph is itself a polygonal graph. We see from Figure 8.10 that, to each face F of the graph G, there corresponds exactly one vertex v^* of the graph G^*, and the number of edges at a vertex v^* of G^* is the same as the number of boundary edges of the corresponding face F in G. This shows that the degree of v^* in G^* is the number of boundary edges of the corresponding face F in G. Each edge e in G corresponds to a unique edge e^* crossing it in G^*.

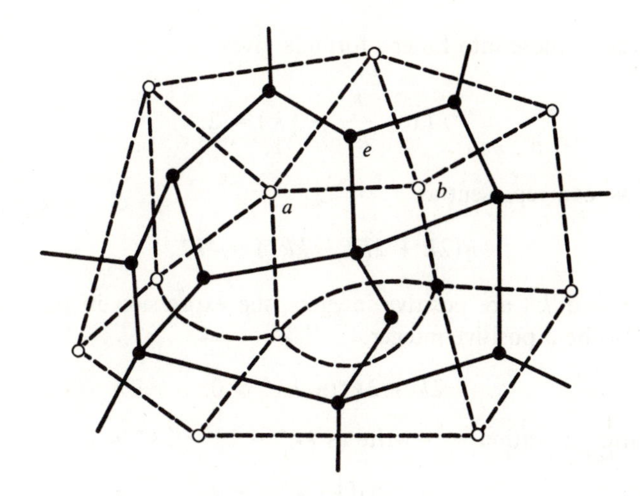

Figure 8.10

At each vertex v in G there are $\deg(v)$ edges. Each of these is crossed by an edge of G^*, and these edges form a face F^* of G^*. Therefore, there are $\deg(v)$ edges of G^* on the boundary of F^*. From the figure, we also see that G is the dual of G^*. The two graphs have the same number of edges, the number of vertices in G^* is the number of faces in G, and the number of faces in G^* is the number of vertices in G.

8.4 The Platonic Solids

We said that a graph is regular (Section 1.6) if the number k of edges is the same at each vertex. For a polygonal graph G, we shall say that it is *completely regular* if also the dual graph G^* is regular. This means (see the preceding section) that each face in G must be bounded by the same number of edges, say k^*.

There are, as we shall show, very few completely regular graphs. If we count the number of edges in G as in the previous section, we find (in the case of completely regular graphs) that these expressions reduce to

$$2m = kn \quad \text{and} \quad 2m = k^*f.$$

Thus

$$m = \tfrac{1}{2}kn \quad \text{and} \quad f = \frac{k}{k^*}n.$$

Substituting these into Euler's formula gives

$$n\left(1 + \frac{k}{k^*} - \tfrac{1}{2}k\right) = 2,$$

which we can represent as

$$n(2k + 2k^* - kk^*) = 4k^*.$$

Since n and k^* are positive integers, the expression in parentheses must also be a positive integer:

$$2k + 2k^* - kk^* > 0.$$

This latter condition we rewrite as $kk^* - 2k - 2k^* < 0$, or

$$(k - 2)(k^* - 2) < 4.$$

We shall solve this inequality in two steps. First, consider the case that both factors, $k - 2$ and $k^* - 2$, are positive; that is, k and k^* are greater than 2. Since the only pairs of positive integers with a product less than 4 are 1 and 1, 1 and 2, and 1 and 3, we see that $k - 2 \leq 3$ and $k^* - 2 \leq 3$. In this case, the only 5 values k and k^* can possibly take are listed in the following table.

Completely regular graphs

k	k^*	n	m	f	type
3	3	4	6	4	tetrahedron
3	4	8	12	6	cube
3	5	20	30	12	dodecahedron
4	3	6	12	8	octahedron
5	3	12	30	20	icosahedron

The number of edges, vertices and faces in the table have been computed from the above expressions for n, m and f. In constructing the completely regular graphs listed in the table, we begin with a triangle, quadrilateral or pentagon, according as the value of k^* is 3, 4 or 5. By fitting the polygons together so that the right number of faces meet at each vertex, we see that there is exactly one isomorphic type (drawn in Figure 8.11) of complete regular graphs for each of the five sets of values.

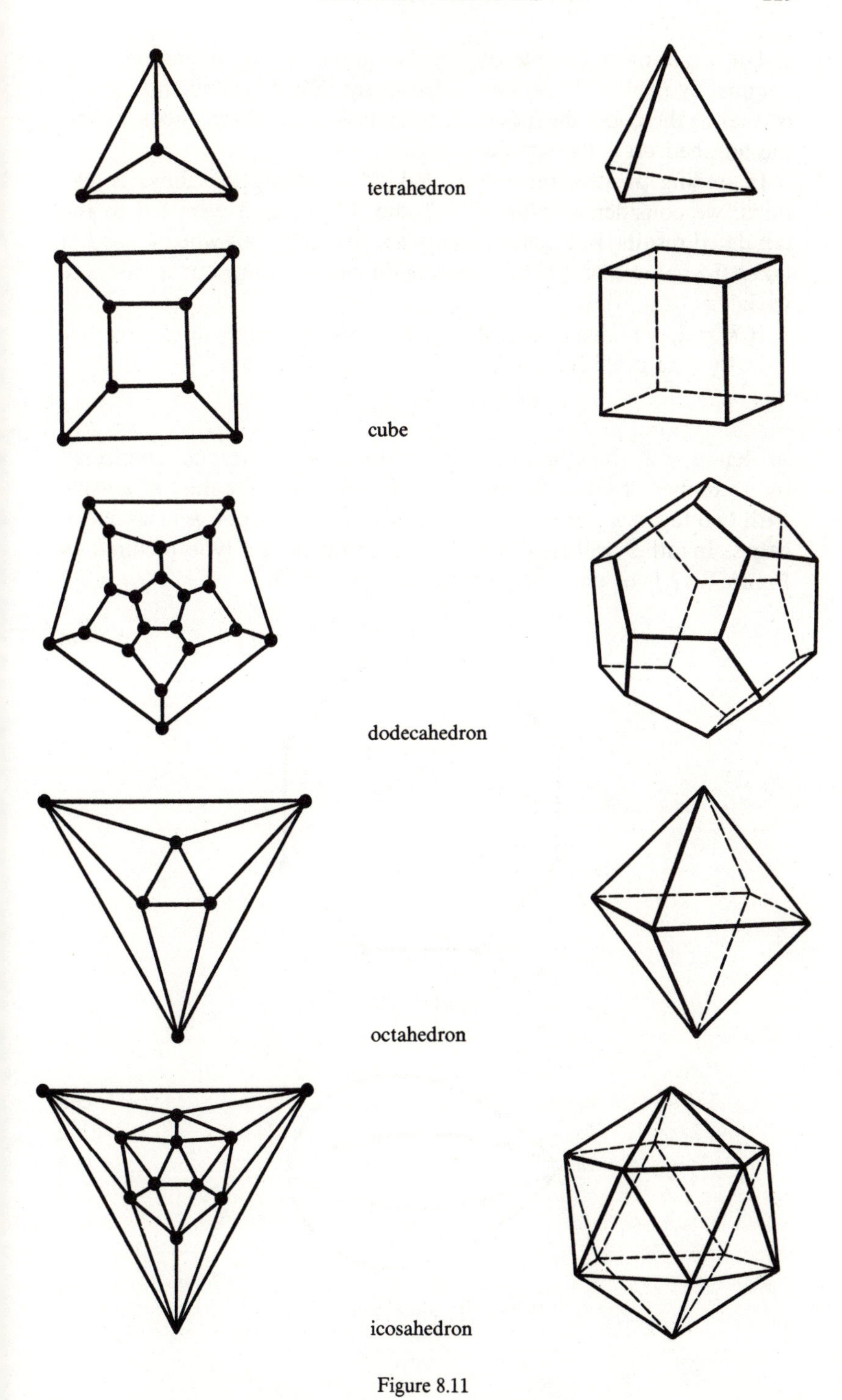

Figure 8.11

The dual of a completely regular graph is, by definition, also completely regular. From our table, we see that the octahedron graph is dual to the cube, the icosahedron is dual to the dodecahedron, and the tetrahedron is its own dual.

In finding positive integers k and k^* satisfying the above conditions, we considered values $k > 2$ and $k^* > 2$ and were led to the tabulated results. But this inequality also has solutions when k (or k^*) takes the value 2 or 1. The corresponding graphs turn out to be quite trivial.

If $k = 2$, we have a connected graph with two edges at each vertex —in short, a cycle (Figure 8.12). If $k^* = 2$, we obtain

$$n(2k + 4 - 2k) = 4n = 8,$$

so that $n = 2$; the graph therefore consists of two vertices connected by a number of edges (Figure 8.13). Observe that the dual of a cycle with two faces, n vertices and n edges has two vertices, n faces and n edges. In other words, the dual of a graph of the type pictured in Figure 8.12 is of the type pictured in Figure 8.13.

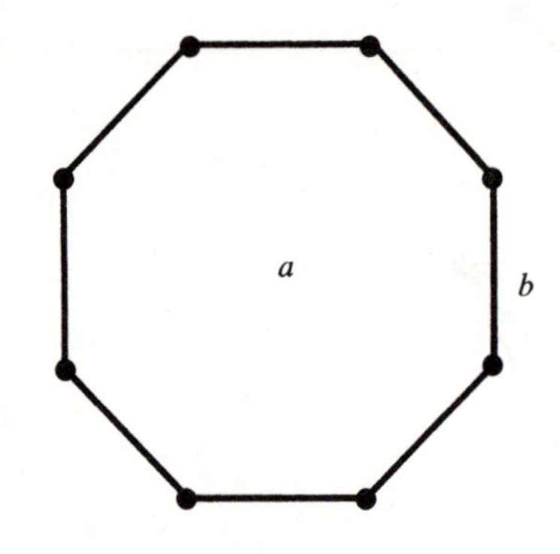

Figure 8.12

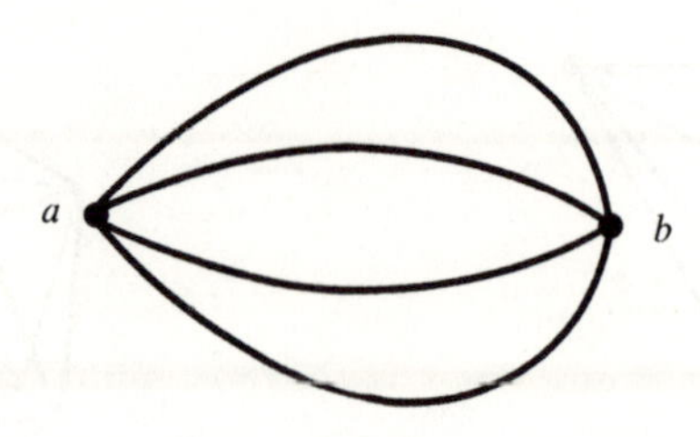

Figure 8.13

When $k = 1$, the above inequalities are satisfied for any positive value of k^*. But a connected graph with only one edge at each vertex must consist of a single edge; that is, it must have

$$n = 2, \quad m = f = 1, \qquad k = 1, \quad k^* = 2.$$

You may verify that, when $k^* = 1$, the graph is a single loop with the dual set of values

$$n = m = 1, \quad f = 2, \qquad k = 2, \quad k^* = 1.$$

In the 13th book of Euclid's *Elements*, there is a discussion of the *regular polyhedra*. These bodies are inscribed in a sphere, all boundary faces are regular and congruent polygons, and at each vertex there are the same numbers of adjoining side edges and faces. The graph of such a polyhedron, defined by its vertices and side edges, is completely regular; it is also planar, because it can be projected upon the sphere from its center.

Plato mentioned the regular polyhedra in *Timaeus*, and such was his influence that they have ever since been known as the Platonic solids. Nor was Euclid the discoverer of these polyhedra; they were known to his predecessors, some even to the Pythagoreans. Throughout antiquity and the Middle Ages, the Platonic solids were considered to be symbols of the harmony of the universe.

It follows from our discussion of completely regular graphs that there can be no Platonic solids other than the five sketched in Figure 8.11.

Problem Set 8.4

1. Draw the duals of the tetrahedron, the cube and the octahedron.
2. Do the regular polyhedra have Hamiltonian cycles? Do they have Eulerian trails?

8.5 Mosaics

When you look upon a bathroom floor, you may see a pattern of regular polygons repeating itself. The shape of the polygons may vary; they may be squares, or triangles, or hexagons (see Figure 8.14).

These patterns for mosaic floors have been popular since antiquity. In nature, we find many instances of such repeated patterns with similar, if not congruent, units. In botany, they are studied under the heading of *phyllotaxis*, the arrangement of buds and seeds in plants; you may be familiar with such regularities from the pattern on a pineapple, or the arrangement of the seeds in a sunflower.

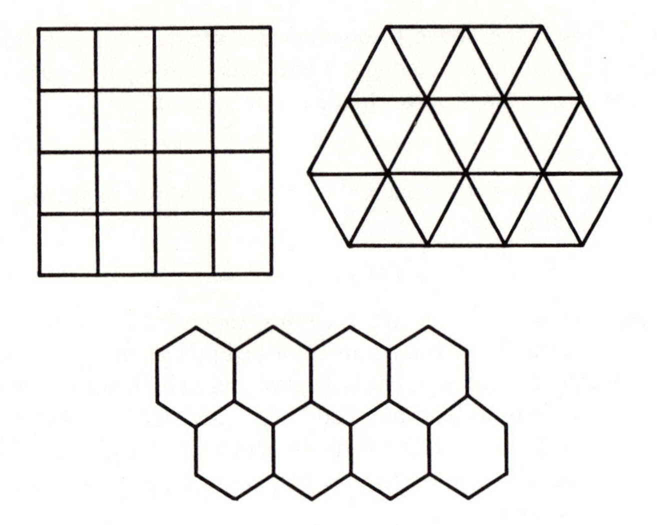

Figure 8.14

From a general point of view, we may say that we have a planar graph in which faces with the same number of edges are repeated a large number of times. We shall show again that there are only a small number of patterns which are actually possible.

Let us use the same notation as before: there are k edges at each vertex, and each face has k^* boundary edges. We begin our mosaic with one of these faces, and add others to it until a certain part of the plane is covered, just as a floor is usually laid.

We now have a polygonal graph G in which all faces except the infinite face are bounded by k^* edges, and there are k edges at each vertex except at those on the boundary of the infinite face.

We suppose that we lay our mosaic in such a manner that, when the number of pieces increases, the proportion of the number b of vertices on the boundary to the total number n of vertices becomes smaller and smaller. In the usual limit terminology, this is expressed symbolically as

$$\frac{b}{n} \to 0, \quad \text{as } n \to \infty;$$

in words: the ratio b/n approaches zero as n approaches infinity.

Let us estimate the number of edges in G by counting them at each vertex, as in the handshaking lemma. If we are generous and count k

edges at each vertex, we obtain kn of them. On the other hand, if we omit the edges at all b vertices on the boundary, we obtain $k(n-b)$ edges. Thus, we know that

$$k(n-b) < 2m < kn,$$

where m is the number of edges in G. We can rewrite these inequalities in the form

$$\frac{k}{2} - \frac{kb}{2n} < \frac{m}{n} < \frac{k}{2}.$$

Since $b/n \to 0$, as $n \to \infty$, we conclude that

$$\frac{m}{n} \to \frac{k}{2}, \quad \text{as } n \to \infty.$$

Next let us count the edges, by means of the faces. There are $f-1$ faces with k^* boundary edges, and the infinite face has b boundary edges, the same as the number of boundary vertices. From this, we conclude that

$$2m = (f-1)k^* + b;$$

we divide both sides of this expression by n and k^*, and then bring it into the form

$$\frac{f}{n} = \frac{2}{k^*}\frac{m}{n} + \frac{1}{n} - \frac{1}{k^*}\frac{b}{n}.$$

When $n \to \infty$, the two last terms on the right tend to zero, and $m/n \to k/2$, and we conclude that

$$\frac{f}{n} \to \frac{2}{k^*}\frac{k}{2} = \frac{k}{k^*}, \quad \text{as } n \to \infty.$$

Let us turn back to Euler's formula, now written in the form

$$1 + \frac{f}{n} = \frac{m}{n} + \frac{2}{n}.$$

For large n, the left-hand side tends to

$$1 + \frac{k}{k^*}$$

while the right-hand side tends to $k/2$. Both sides must tend to the same limit, and we conclude that

$$1 + \frac{k}{k^*} = \frac{k}{2},$$

and this condition we finally rewrite as

$$(k - 2)(k^* - 2) = 4.$$

The only pairs of integers that can satisfy this equation are

$$k = 3, \quad k^* = 6; \qquad k = 4, \quad k^* = 4; \qquad k = 6, \quad k^* = 3.$$

We conclude that all repetitive planar graph patterns or mosaics must be formed either by triangles, or by quadrilaterals, or by hexagons. All three of these are illustrated in Figure 8.14.

CHAPTER NINE

Map Coloring

9.1 The Four Color Problem

When we have a polygonal map before us, we may think of the faces as being countries or states on a map, with the ocean surrounding them in the form of the infinite face. In a good atlas the countries, together with the ocean, are colored in different colors to distinguish them from each other. This means that the coloring must be done so that countries with a common boundary have different colors. If one has a large number of colors at one's disposal, this represents no particular problem. Much more difficult is the question of finding the smallest number of colors sufficient for coloring the countries of a given map.

A famous problem is to prove that every map can be colored properly by means of four colors. The earliest trace of it we find in a letter of 23 October 1852 from the London mathematician Augustus De Morgan to Sir William Rowan Hamilton in Dublin: "A student of mine asked me today to give him a reason for a fact which I did not know was a fact and do not yet. He says that if a figure be anyhow divided and the compartments differently coloured, so that figures with any portion of common boundary line are differently coloured—four colours may be wanted but no more." The student was Frederick Guthrie, later a physicist, who stated that he learned the problem from his brother Francis Guthrie, later a mathematician, who had guessed it in connection with a coloring of a map of England.

At first the problem does not seem to have been taken too seriously; mathematicians appear to have considered it a fairly self-evident fact. Later, a number of incorrect proofs appeared, and for many years the four color problem, so puzzlingly simple to state, withstood every assault by some of the world's most capable mathematicians. However, interest in graph theory was mightily stimulated by the problem; many important graph results were discovered because they showed promise of being helpful for the resolution of the four color problem. It was not until 1976 that the four colour problem was finally settled, and that four colours were proved to be sufficient in all cases. The solution, due to K. Appel and W. Haken of the University of Illinois, was exceedingly complicated, and made extensive use of a computer. A brief discussion of their proof appears at the end of Section 9.2, but we can state their result here:

FOUR COLOR THEOREM. *Every map can be colored with four colors.*

Let us point out immediately that, for some maps, it is essential that we have as many as four colors. As an example, we may take the case of the tetrahedron map (Figure 9.1). (In the following we shall always denote the various colors by Greek letters, $\alpha, \beta, \gamma, \delta, \epsilon, \ldots$; usually it does not matter which particular colors the letters indicate.) For the tetrahedron, we may suppose that the infinite face has the color α. The other three faces must then have different colors, since they all have common boundaries.

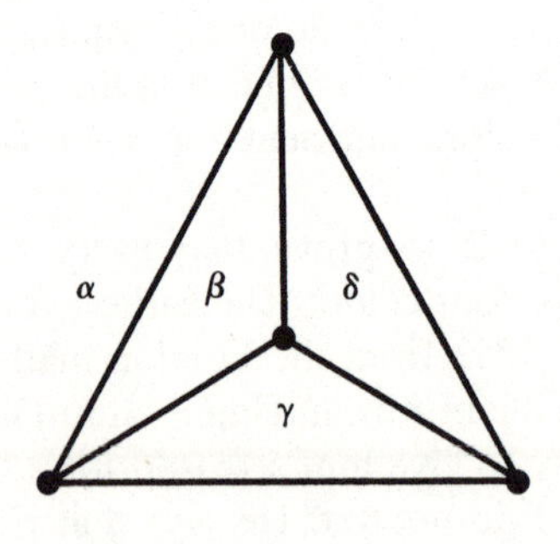

Figure 9.1

We observe next that in trying to color a map with as few colors as possible, we need not concern ourselves with vertices where only two edges meet. For, if a is such a vertex, we can combine the two edges into one and eliminate the vertex a (Figure 9.2) without changing the color scheme in any way.

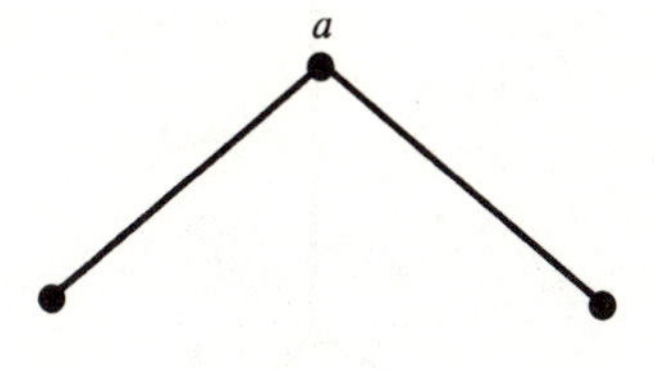

Figure 9.2

According to this remark, it suffices to consider maps where there are at least three edges at each vertex. But we can make a much more radical supposition: the coloring of an arbitrary polygonal graph in a certain number of colors can be reduced to the case where the graph is *regular of degree three*—that is, there are *exactly* three edges at each vertex. In other words, if we can solve the coloring problem for regular polygonal graphs of degree 3, then we can solve it for all maps.

Suppose that we have more than three edges at some vertex a (Figure 9.3). We draw around a a small circle C which does not reach any other vertex. We then eliminate a and those parts of its edges which lie inside the circle, and replace them by edges consisting of the sections of C (Figure 9.4). All the new vertices will have three edges. Any coloring of the new graph will produce a coloring for the original graph when we shrink the circle C down to a single point. By repeating this process for each vertex with more than three edges, we

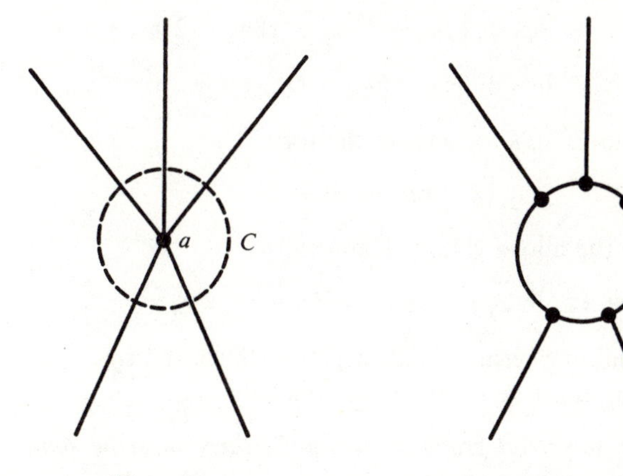

Figure 9.3

Figure 9.4

reduce our coloring problem to that of coloring a regular polygonal graph of degree three.

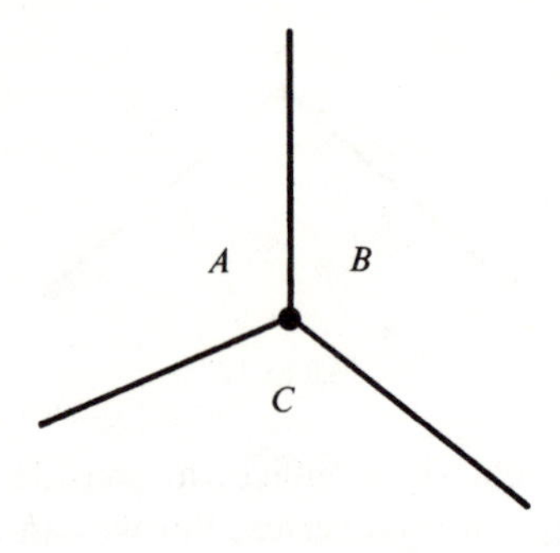

Figure 9.5

In the following, we shall restrict ourselves to regular graphs of degree three. For such graphs, there is a useful formula in addition to those derived in Section 8.3; it is based on the observation that each vertex now lies on the boundary of exactly three faces (Figure 9.5). As a consequence, when we count the number of vertices on all faces, we obtain three times the total number of vertices. Previously we denoted by φ_i the number of faces with i edges and i vertices. Thus,

$$3n = 2\varphi_2 + 3\varphi_3 + 4\varphi_4 + 5\varphi_5 + 6\varphi_6 + 7\varphi_7 + \cdots .$$

If we multiply this expression by 2, and multiply the equations in Section 8.3 for m by 3, and for f by 6, we obtain the expressions

$$6n = 4\varphi_2 + 6\varphi_3 + 8\varphi_4 + 10\varphi_5 + 12\varphi_6 + 14\varphi_7 + \cdots;$$

$$6m = 6\varphi_2 + 9\varphi_3 + 12\varphi_4 + 15\varphi_5 + 18\varphi_6 + 21\varphi_7 + \cdots;$$

$$6f = 6\varphi_2 + 6\varphi_3 + 6\varphi_4 + 6\varphi_5 + 6\varphi_6 + 6\varphi_7 + \cdots .$$

We now rewrite Euler's formula in the form

$$12 = 6n - 6m + 6f,$$

and substitute the above values. The result is

$$12 = 4\varphi_2 + 3\varphi_3 + 2\varphi_4 + \varphi_5 - \varphi_7 - 2\varphi_8 - \cdots,$$

where the remaining terms are all negative. Since the right-hand side of this equation must be positive, we conclude:

in a regular polygonal graph of degree 3, *there must be some faces bounded by less than six edges.*

9.2 The Five Color Theorem

In the following, we shall examine the possibility of coloring a map in four or five colors; according to the preceding remarks, we can assume that we deal with a polygonal regular graph G of degree 3, and we know that such a graph has at least one face bounded by fewer than six edges.

In what follows, we shall deal separately with faces bounded by 2, 3, 4 and 5 edges. In each case we shall show that

(a) some boundaries can be deleted, so that the resulting graph is still regular of degree three but has fewer faces;

(b) if the reduced graph can be colored with no more than five colors, then this can be done for the original graph as well. Since our reductions always lead to regular graphs of degree three, we can be sure that, after each such simplification, there is again a face bounded by fewer than six edges; we are therefore led, successively, to graphs with fewer and fewer regions to be colored.

Multiple edges

(a) Our first step is to show that G can always be simplified so that there are no faces bounded by two edges. If there is a double edge between a and b, as in Figure 9.6, then we eliminate one of them, and replace the three edges ca, ab and bd by a single edge cd. The new graph G_1 is still regular of degree 3.

(b) If G_1 can be colored, there will be different colors α and β on the two sides of the edge cd. We can then restore a and b and the double edge, and give the enclosed face A a third color γ.

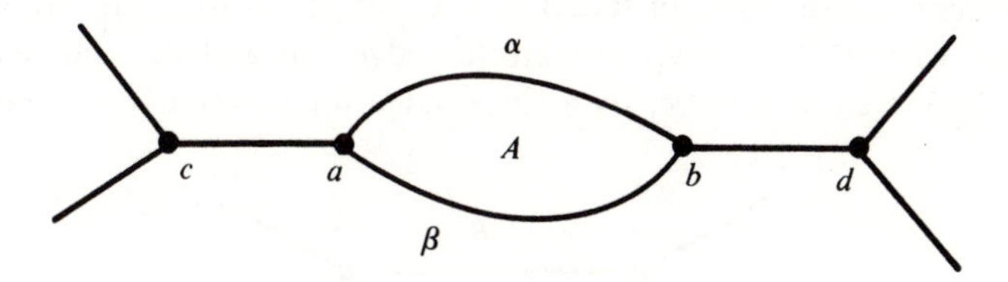

Figure 9.6

Triangular faces

(a) The graph G can be simplified so that it includes no triangular faces. Suppose that D is such a face bordering upon three others A,

B, C. Let ab be the boundary edge between C and D, as in Figure 9.7. We eliminate the edge ab and combine the other two edges at a into a single one, and similarly at b. The reduced graph G_1 is then regular of degree 3.

(b) When G_1 has been colored, the face A will have a color α, B a color β, and the face $C + D$ a color γ. When the edge ab is restored, we need only give D a fourth color δ.

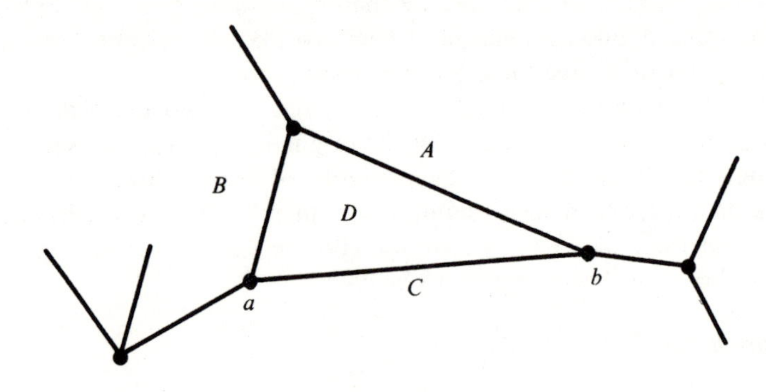

Figure 9.7

Quadrilateral faces

(a) It is a little more involved to show that we can eliminate faces with four edges on the border. Let F be such a face with neighboring faces A, B, C, D, as in Figure 9.8. It is possible that A and C are only different parts of the same face, or it may happen that they have a boundary edge mn in common, as we have indicated in Figure 9.9. In either case, the face (or faces) $A + C$ cuts B off from any common border with D. We now eliminate the edges ab and cd, and replace the edges a_1a, ac and cc_1 by a single edge a_1c_1; similarly, we replace

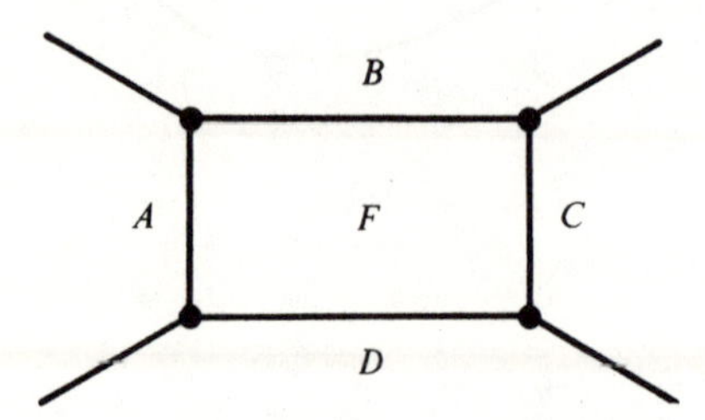

Figure 9.8

the edges b_1b, bd and dd_1 by a single edge b_1d_1. Then the new graph G_1 is regular of degree 3, and $B + F + D$ is a single face.

(b) Suppose that G_1 has been colored, and $B + F + D$ has the color α. Then A and C have colors β and γ, possibly alike. We can then restore the two edges ab and cd, and give F a fourth color δ.

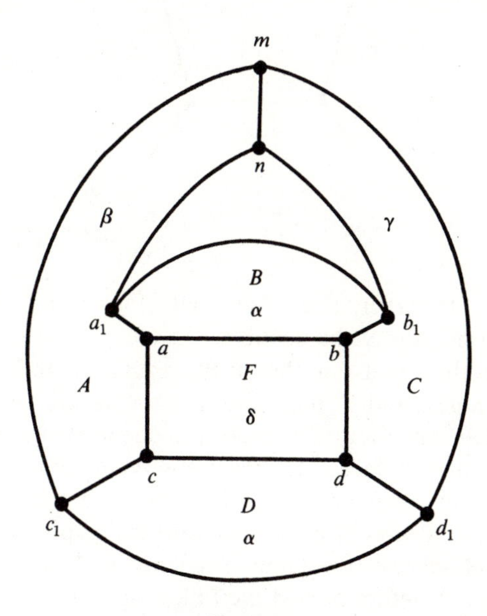

Figure 9.9

Pentagonal faces

(a) Suppose that we have a face F bordering upon five others, A, B, C, D, E, as in Figure 9.10. By the same argument as that given for the quadrilateral, we conclude that there is a pair of oppositely located bordering faces to F, say A and C, which are not parts of the same face and do not have a common border. We now eliminate the edges ab and cd (see Figure 9.10). As before, the graph becomes regular of degree 3, if we eliminate the vertices a, b, c and d by joining the two remaining edges to a single edge at each of these vertices.

(b) Suppose that we have colored this reduced graph, and that the face $A + F + C$ has the color α. The three faces B, D and E may require up to three different colors, β, γ, δ. If we have five colors at our disposal, then we can restore the two edges ab and cd, and give F

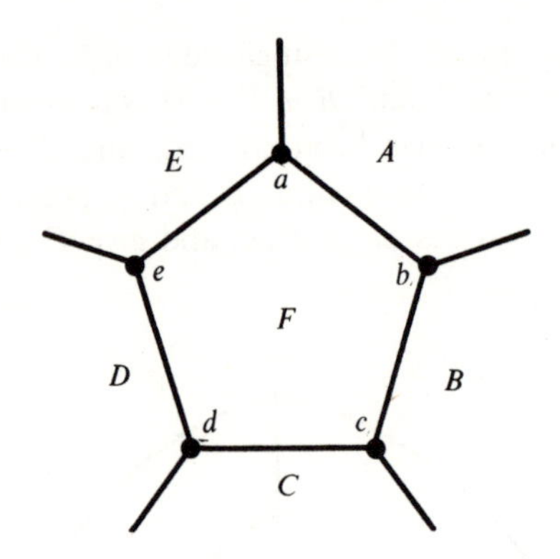

Figure 9.10

a fifth color ε. However, if we have only four colors, then such a reduction may not always be possible. □

We see from this discussion that, if the regular graph has faces with 2, 3, 4 or 5 edges, and if there are five colors available, then the coloring problem can always be reduced to one involving a graph with a smaller number of faces. The last observation in the preceding section shows that a regular graph has at least one such small face, so the reduction can always be continued until the graph has only five or fewer faces, and hence can certainly be colored with no more than five colors. We have therefore proved the following result:

FIVE COLOR THEOREM. *Every map can be colored with five colors.*

This argument is not applicable when we have only four colors. As we saw, pentagons cannot be reduced in this case. Our reduction process may come to a halt with some regular graph where each face has at least five boundary edges. Then there are no faces with 2, 3, or 4 boundary edges, and we have

$$\varphi_2 = \varphi_3 = \varphi_4 = 0.$$

The last equation in Section 9.1 shows that such an irreducible regular graph must have at least 12 pentagons. In fact, it can also be proved that every irreducible regular graph must have either two neighboring pentagons or a pentagon joined to a hexagon. We refer to this pair of configurations as an *unavoidable set*, since every irreducible regular graph must contain at least one of them.

Many unavoidable sets have been constructed. In their proof of the four color theorem, Appel and Haken constructed an unavoidable set with almost 2000 configurations. Each of these configurations was examined individually, using 'reduction procedures' similar to (but much more complicated than) those used above for multiple edges, triangular faces, quadrilateral faces, and pentagonal faces. Since every irreducible regular graph must contain at least one of these 2000 configurations, and since all of them can be dealt with on an individual basis, it follows that the four color theorem is proved: every map can indeed be colored with just four colors.

9.3 Coloring Maps on Other Surfaces

The four color problem is not the only difficult problem to have been settled recently. A related problem, which is sometimes called the *Heawood Problem*, was also unsolved for many years, until it was finally settled in 1968 by G. Ringel and J.W.T. Youngs.

You will recall that the four color problem can also be expressed as a result about the coloring of maps on a sphere:

FOUR COLOR THEOREM. *Every map drawn on the surface of a sphere can be colored with four colors.*

It is natural to ask how many colors are needed to color maps on some other surfaces. For instance, how many colors are needed to color maps on a ring-shaped surface (Figure 9.11), known to mathematicians as a *torus*?

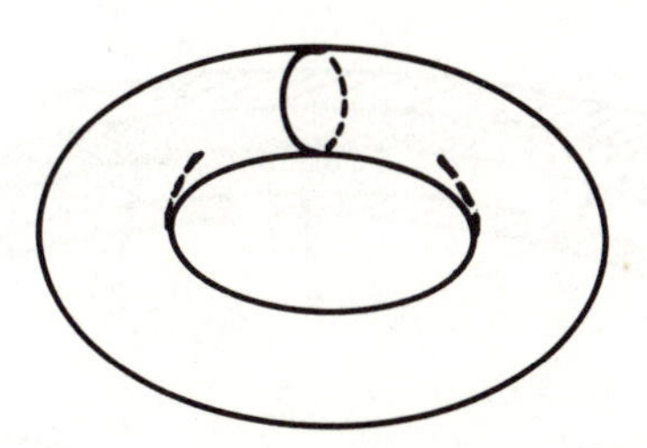

Figure 9.11

When drawing maps on the surface of a torus, we now have the extra flexibility of being able to draw edges around the ring, or through the hole in the center. For instance, we saw in Section 8.1 that neither the houses-and-wells graph (Figure 8.1) nor the complete graph K_5 (Figure 8.2) can be drawn without crossings on the surface of a sphere, but both can be drawn on the surface of a torus (see Figure 9.12).

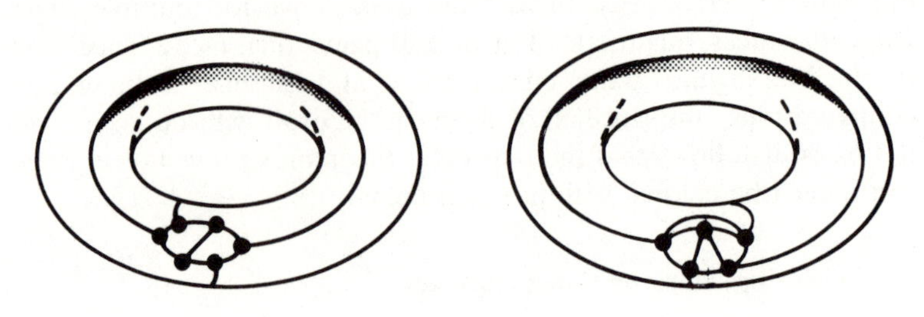

Figure 9.12

Many of the standard results for planar graphs have their analogues for graphs drawn on a torus; for instance, there is a version of Euler's formula for such graphs, and Neil Robertson and Paul Seymour have recently obtained an analogue of Kuratowski's theorem (see Section 8.1). In particular, it is not difficult to obtain the following analogue of the four color theorem:

SEVEN COLOR THEOREM. *Every map drawn on the surface of a torus can be colored with seven colors.*

Just as there are maps on the sphere which require all four colors, so there are maps on the torus which require all seven colors. An example of such a map is shown in Figure 9.13.

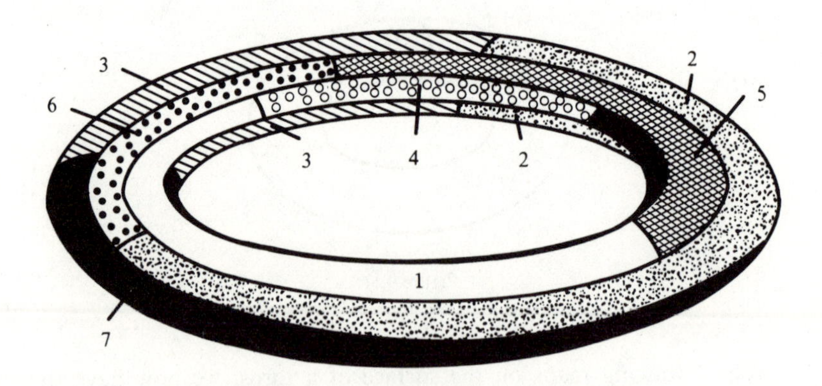

Figure 9.13

We can extend this discussion and ask how many colors are needed to color maps drawn on ring-shaped surfaces with extra holes, such as the two-holed ring shown in Figure 9.14. A formula for the number of colors needed for map-coloring on such surfaces was derived by the English mathematician Percy Heawood (1861–1955):

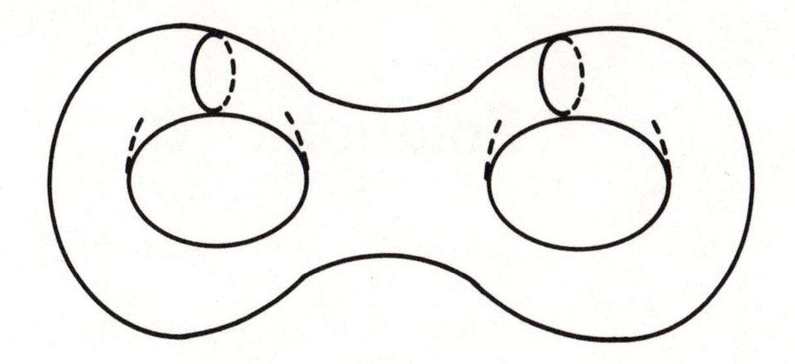

Figure 9.14

HEAWOOD'S FORMULA. *Every map drawn on a surface with g holes can be colored with* $[\frac{1}{2}(7 + \sqrt{1 + 48g})]$ *colors.*

In this formula, $[x]$ denotes the integer part of x (so that, for instance, $[3] = [\pi] = 3$); note that, when $g = 1$, this formula gives the correct answer, 7.

What Heawood did not prove, however, was that there are maps which actually require this number of colors, just as there are maps on a torus which need 7 colors. Surprisingly, proving this turned out to be a major problem, which became known as the Heawood problem. It was only after a major struggle, lasting almost eighty years, and involving ideas from the theory of electrical networks, that a proof was finally obtained.

Solutions

Chapter 1

Problem Set 1.1

2. *AB*, *AD*, *AE*, *BC*, *BF*, *BG*, *CD*, *CG*, *DH*, *EF*, *EG*, *EH*, *FG*, *FH*, *GH*.
3. 6 vertices, 9 edges; 8 vertices, 15 edges.

Problem Set 1.2

1.

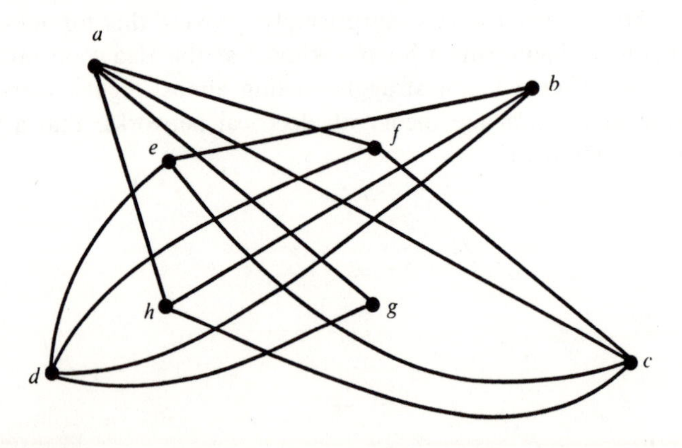

2. 10; 15; 21.
3. $\frac{1}{2}n(n-1)$

Problem Set 1.3

1. Graph 1.1 is not isomorphic to graph 1.2, for the latter has eight vertices, while the former has just six. Also, graph 1.1 is not isomorphic to graph

1.6, for the latter has a vertex f with only one edge; for the same reason, graphs 1.2 and 1.6 are not isomorphic.

2. The left-hand edge of the first graph belongs to two cycles of length 4, but there is no such edge in the second graph.
3. When the vertices correspond as in the figure then the edges correspond.

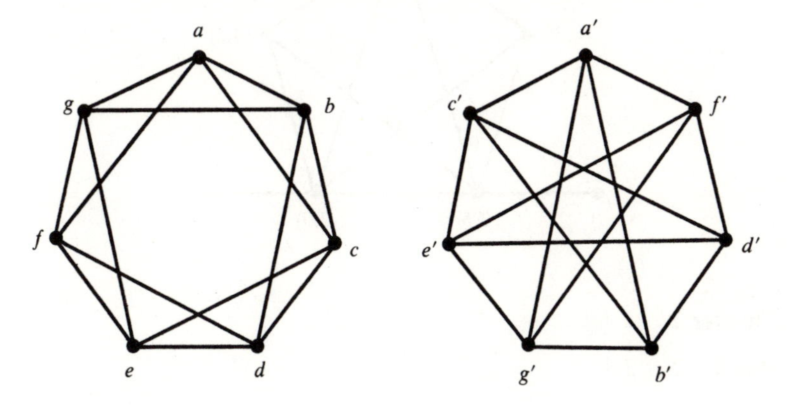

Problem Set 1.5

1. If the four original neighbors are a, b, c, d, their roads form a planar graph as in the figure below.

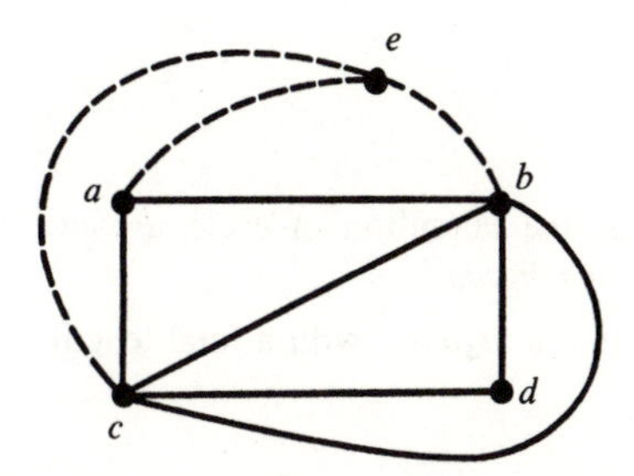

2. Wherever the fifth point e is placed, it is always separated from one of the others by a closed cycle; in the figure, it cannot be joined to d.

Problem Set 1.6

1. In Figure 1.2, $\deg(a) = \deg(c) = \deg(d) = 3$, $\deg(b) = \deg(e) = \deg(f) = \deg(h) = 4$, $\deg(g) = 5$, and there are $\frac{1}{2}\{(3 \times 3) + (4 \times 4) + 5\} = 15$ edges. In Figure 1.6, we have $\deg(a) = \deg(b) = \deg(d) = \deg(e) = 2$, $\deg(c) = 3$, $\deg(f) = 1$, and there are $\frac{1}{2}\{(4 \times 2) + 3 + 1\} = 6$ edges.
2. There are, respectively, 4 and 2 odd vertices.

Problem Set 1.7

1.

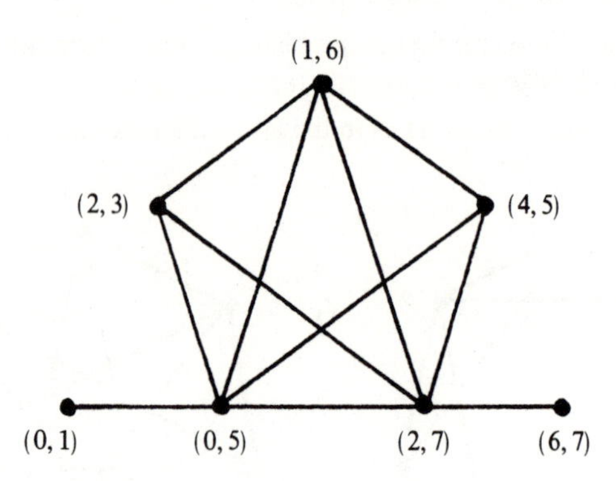

Chapter 2

Problem Set 2.3

1. There are two odd vertices in each graph in Figure 2.7; hence each graph can be covered in a single path.
3. The graph K_4 can be covered by the two paths $1,2,3,4$ and $2,4,1,3$. For K_5, take the single path $1,2,3,4,5,1,4,2,5,3,1$. In general, K_n can be covered by a single path only when n is odd; when n is even, $\frac{1}{2}n$ paths are necessary.

Problem Set 2.5

1. The first graph has the Hamiltonian cycle *acbfeda*; the second has the Hamiltonian cycle *abfehgcda*.
2. The shortest path is $a_1a_2a_4a_3a_1$, with a total length of 470 miles.

Problem Set 2.6

2. The number of persons on the second side increases by at most one person at each passage. Thus, if the problem were solvable, there would be at one stage 5 persons transferred. These cannot consist of 4 women and 1 man, nor of 3 women and 2 men, because one woman would be without her husband. Nor can there be 2 women and 3 men, because on the other side of the river a woman would be without her husband. The 5 persons transferred must therefore be 1 woman and 4 men. In the last transfer, 1 woman, 1 man or 2 men must have arrived. But both are impossible, since in the first case one man must have been alone with 4 women on the first side, and in the second, 2 men would have been there with 3 women. If the ferryboat holds 3 persons, and the pairs are Aa, Bb, Cc, Dd, then the

transfers can be made in the following stages: (a, b, c), (a, b, c, d), (Aa, Bb, Cc), (Aa, Bb, Cc, D), (Aa, Bb, Cc, Dd).

3. 336 moves when directions are taken into account, 168 otherwise.
4. There are 4 corner positions with 3 moves, 24 side positions with 5 moves, and 36 central positions with 8 moves, giving a total of 420 directed moves, or 210 undirected ones.

Chapter 3

Problem Set 3.2

1. The cycle ranks of these graphs are, respectively,

$$\gamma = 15 - 8 + 1 = 8 \quad \text{and} \quad \gamma = 32 - 21 + 1 = 12.$$

2. $\gamma = \frac{1}{2}(n-1)n - n + 1 = \frac{1}{2}(n-1)(n-2)$.

Problem Set 3.3

1. The trees with edges *bd*, *ab*, *ac*, and *bd*, *ad*, *ac*.

Problem Set 3.4

1. (i) 23; (ii) 23; (iii) 24.

Problem Set 3.5

1. Framework (i) is not rigid. Framework (ii) is rigid, and the bracing is a minimum bracing.
2. $m + n - 1$.

Problem Set 3.6

1. For the graph in Figure 1.1, we have, for instance:

$$a \to ac, \quad b \to be, \quad c \to cb, \quad d \to da, \quad e \to ef, \quad f \to fa.$$

For the graph in Figure 1.2, we have:

$$a \to ab, \quad b \to bc, \quad c \to cd, \quad d \to da,$$
$$e \to ef, \quad f \to fg, \quad g \to gh, \quad h \to he.$$

Chapter 4

Problem Set 4.2

1. A trivial example consists in letting each member be a committee of one. Then no one could serve as secretary of any other committee.

2. The number of committees is

$$\frac{12 \cdot 11 \cdot 10}{1 \cdot 2 \cdot 3} = 220;$$

this is much larger than the number 12 of possible secretaries, so it would not be possible to assign separate secretaries to each of them.

3. One matching is $m_1 \to p_2$, $m_2 \to p_1$, $m_3 \to p_4$, $m_4 \to p_5$.

Problem Set 4.3

2. Properties (a) and (b) are consequences of the fact that each line and column in Figure 4.7 includes every number from 1 to n exactly once. To prove statement (c), we analyze the construction of the table in Figure 4.7. If we count also the first row and the first column, then there are n rows and n columns whose entries are constructed by the following rule:
   (i) In the ith row and the ith column, player i faces player 1.
   (ii) When $i \neq j$, player j faces, in the ith row, the player whose number is

$$k = 2i - j\,[\pm(n-1)]$$

(where we add or subtract $n - 1$, if necessary, to make k lie between 2 and n). If we solve this expression for j, we obtain

$$j = 2i - k\,[\pm(n-1)];$$

the symmetric roles played by k and j confirm the assertion made in part (c).

Chapter 5

Problem Set 5.3

2. We must write

$$\text{outdeg}(a) + \text{indeg}(a) + \text{deg}'(a) = k,$$
$$\text{outdeg}(b) + \text{indeg}(b) + \text{deg}'(b) = k - 1,$$

where deg′ refers to the number of draws.

3. For $n = 5$, we have the graph

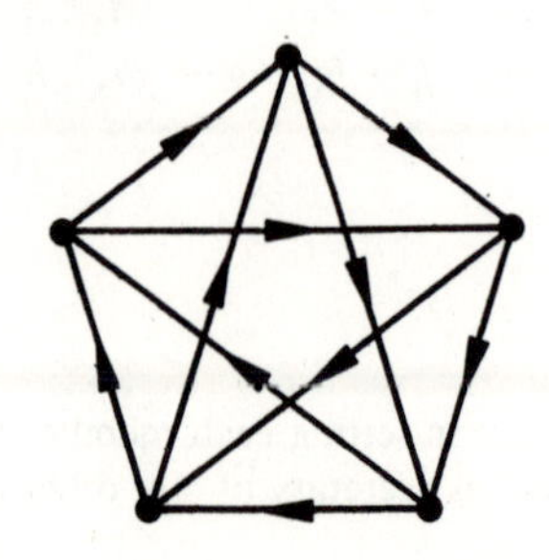

Problem Set 5.4

1. The graphs are

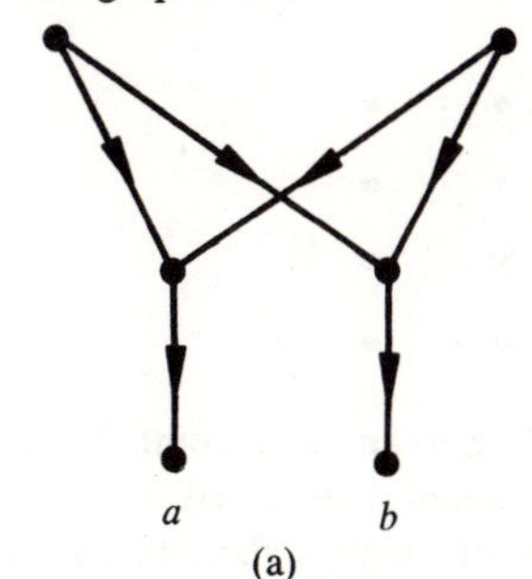

(a)

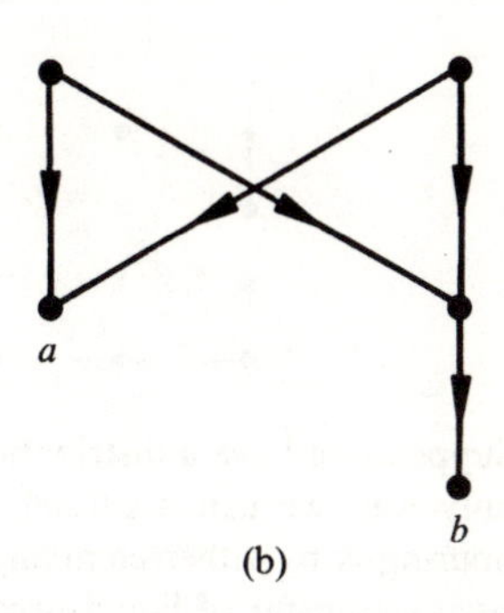

(b)

2. If 1 is taken to be male, then 2 and 7 are females, 5 is male, and the other vertices can be either male or female.
3.

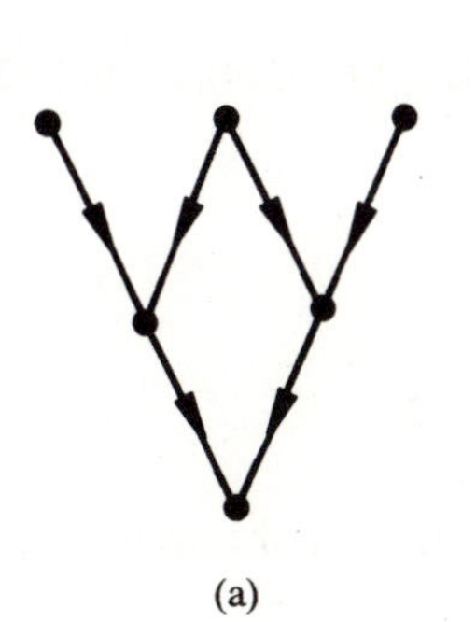

(a)

(b)

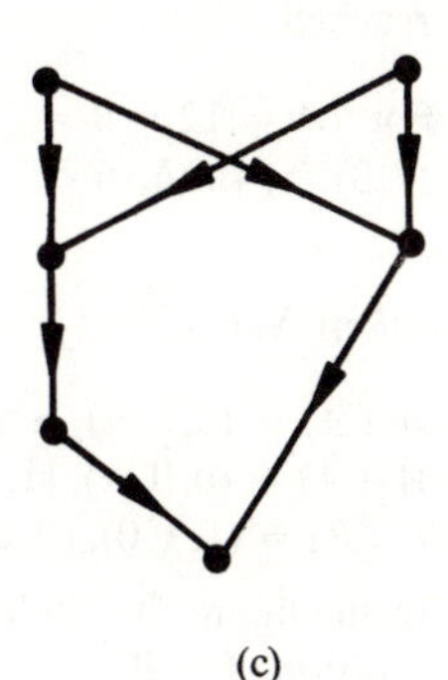

(c)

Problem Set 5.5

1. The shortest time to reach city t is 10, and the shortest route is *sbct*.

Chapter 6

Problem Set 6.1

1. Since the only measuring devices at our disposal are the jugs B and C themselves, we have no way of judging how the liquid is distributed except when one of these jugs is either full or empty. All such distributions of the liquid are represented by points on the border of the rectangle in the figure

below. The possible distributions are represented by the inner points (marked by x) of this rectangle.

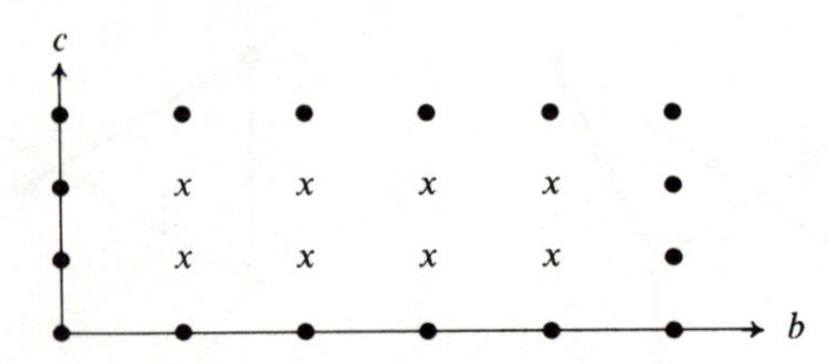

2. Suppose we have a distribution corresponding to an inner point. Then the only way we can accurately measure the liquid transferred in the next pouring is by either emptying or filling one of the jugs B, C. In any case, such a transfer of liquid necessarily leads to a point on the border of the graph. Once we have a full or an empty jug, the next measurable transfer of liquid leads again to a distribution where at least one of the jugs is either full or empty. In other words, each measurable pouring from a border position leads again to a border position, so the inner points u cannot be reached.

3. For $A = 12$, $B = 7$, $C = 4$, a solution is $(7,0), (3,4), (3,0), (0,3), (7,3), (6,4), (6,0)$.

Problem Set 6.2

2. $W_1(A) = (0,0,\alpha),\ \alpha \geq 1;\ L_1(B) = (1,1,0)$;
   $W_2(A) = (0,0,\alpha),\ (1,1,\beta),\ (0,1,\gamma),\ \beta \geq 1,\ \gamma \geq 2$;
   $L_2(B) = (1,1,0),\ (0,2,2)$.

3. In the figure the circles indicate winning positions for A, the squares losing positions for B.

4	□				
3	○	○			
2	□	○	□		
1	○	○	○	○	
0	□	○	□	○	□
	0	1	2	3	4

$W_1(A) = (0,1), (1,1), (1,0)$

$L_1(B) = (0,2), (2,0)$

$W_2(A) = (3,0), (3,1), (2,1), (1,2), (0,3), (1,3), (0,1), (1,1), (1,0), \ldots$

$L_2(B) = (2,2), (4,0), (0,4), (0,2), (2,0), \ldots$

Problem Set 6.3

2. The draws must be represented as edges in both directions.

Chapter 7

Problem Set 7.1

1. Take, for example, the relations "a is greater than the square of b," or "the sets A and B have no common elements", or "one triangle intersects another".

2. (a)

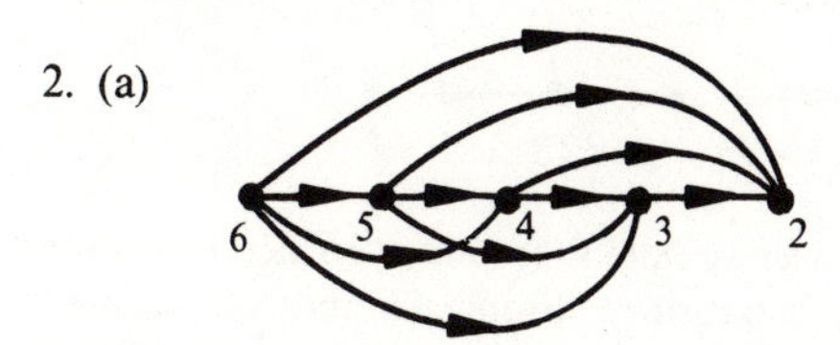

R_6: {2,3,4,5}, R_5: {2,3,4}, R_4: {2,3}, R_3: {2}, R_2 contains no elements

(b) This relation is represented by the complete graph with 5 vertices (without loops); R_6: {2, 3, 4, 5}, R_5: {2, 3, 4, 6}, R_4: {2, 3, 5, 6}, R_3: {2,4,5,6}, R_2: {3,4,5,6}.

(c)

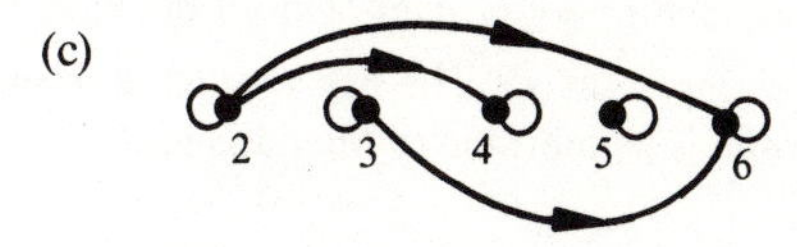

R_6: {6}, R_5: {5}, R_4: {4}, R_3: {3,6}, R_2: {4,6}.

Here we have interpreted $x|y$ to be reflexive—that is, every number divides itself; see also the solution to Problem 3 of Problem Set 7.4.

Problem Set 7.2

2. (a) This relation is anti-reflexive, asymmetric and transitive.
   (b) This relation is reflexive, symmetric, but not transitive.

Problem Set 7.3

2. If $a = b + km$ and $c = d + k'm$, then

$$a + c = b + d + (k + k')m,$$
$$a - c = b - d + (k - k')m.$$

Therefore, $a \pm c \equiv b \pm d (\text{mod } m)$. Also,

$$ac = bd + (dk + bk')m + kk'm^2$$
$$= bd + (dk + bk' + kk'm)m.$$

Therefore, $ac \equiv bd (\text{mod } m)$.

3. The relation $|a| = |b|$, for numbers a, b, is reflexive, symmetric, and transitive, so it is an equivalence relation. Each equivalence class consists of the pair $(a, -a)$, and the members of each pair are distinct except when $a = 0$.

Problem Set 7.4

1. In the complete order graph, the edges are

$$(4,3),(4,2),(4,1),(3,2),(3,1),(2,1);$$

the basis graph is the directed path $4 \to 3 \to 2 \to 1$.

2. The basis graph is the directed path

3. To show that $a|b$ is a partial order, we note first that $a|a$ since $a = 1 \cdot a$. Next, we verify that the relation is transitive: From $a|b$ and $b|c$, we have $b = ka$ and $c = k'b$ so that $c = k'ka$, and hence $a|c$. Finally, if $a|b$ and $b|a$, then $b = ka$, $a = lb$, so $b = klb$; this can hold only if the integers k and l have the value 1—that is, if $a = b$. For a strict partial order, we require for $a|b$ that $b = k \cdot a$ with $k \neq 1$; in other words, if a must be a proper divisor of b in the relation $a|b$, then we have an anti-reflexive relation. In the strict partial order, $a|b$ and $b|a$ cannot both be allowed.

4. There are 8 subsets, including the empty set $\varnothing$. The basis graph is drawn here; observe that the converse graph is isomorphic to the basis graph.

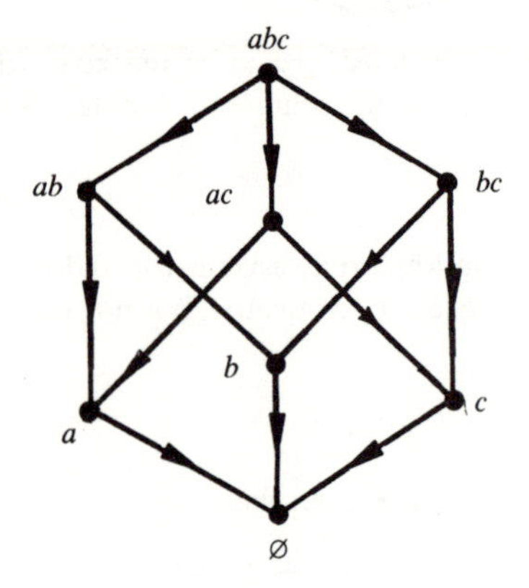

Chapter 8

Problem Set 8.1

1.

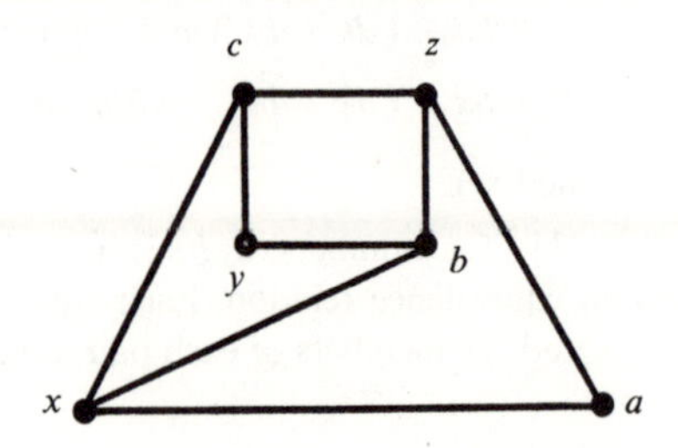

2. They consist of the graph in Figure 8.1, the graph obtained from Figure 8.2 by adding a vertex on one of the edges, and all graphs obtained from these by adding further edges.

Problem Set 8.2

1. $n = 7$, $m = 11$, $f = 6$, and $7 - 11 + 6 = 2$.
2. In general, $n = (f + 1)^2$, $m = 2k(k + 1)$, $f = k^2 + 1$, and $n - m + f = 2$.

Problem Set 8.4

2. All the regular polyhedra have Hamiltonian cycles. Only the octahedron has an Eulerian trail.

BIBLIOGRAPHY

There are a number of good books on graph theory. An introductory book, at about the same level as this one, is

G. Chartrand, *Introductory Graph Theory*, Dover, New York, 1985.

A more substantial book, with a wide range of applications, is

R. J. Wilson and J. J. Watkins, *Graphs: An Introductory Approach*, John Wiley & Sons, New York, 1990.

An introductory text, at a slightly higher mathematical level, is

R. J. Wilson, *Introduction to Graph Theory*, 3d. edn., Longman, Harlow, Essex, 1985.

A classic text, by the same author as this one, is

O. Ore, *Theory of Graphs*, American Math. Soc. Colloq. Publ. XXXVIII, Providence, Rhode Island, 1962.

Other standard texts in graph theory are

J. A. Bondy and U. S. R. Murty, *Graph Theory with Applications*, American Elsevier, New York, 1979.

G. Chartrand and L. Lesniak, *Graphs and Digraphs*, 2d. edn., Wadsworth & Brooks/Cole, Monterey, California, 1986.

F. Harary, *Graph Theory*, Addison-Wesley, Reading, Mass., 1969.

C. Berge, *Graphs*, North-Holland, Amsterdam-New York, 1985.

A historical approach to graph theory, including edited versions of many of the original papers in the subject, can be found in

N. L. Biggs, E. K. Lloyd and R. J. Wilson, *Graph Theory 1736–1936* (paperback edn.), Clarendon Press, Oxford, 1986.

A considerable number of applications of graphs to puzzles and games can be found in

W. W. Rouse Ball, *Mathematical Recreations and Essays*, London, 1892, and Macmillan, London, 1962.

Finally, extended accounts of the four color problem and related material can be found in

T. L. Saaty and P. C. Kainen, *The Four-Color Problem* (2d. edn.), Dover, New York, 1986.

K. Appel and W. Haken, *Every Planar Map is Four Colorable*, Contemporary Mathematics, Volume 98, American Mathematical Society, 1989.

GLOSSARY

bridge: an edge whose removal would increase the number of connected components of the graph.

complement $\overline{G}$ of a graph G: $\overline{G}$ consists of the vertices of G and the edges which are not in G.

connected component of a vertex a: the set of all vertices reachable from a by paths in the graph and all edges incident on these vertices.

cycle: a path that returns to its starting point—that is, a route that revisits only the beginning vertex.

maximal cycle of a polygonal graph G: the cycle which surrounds the whole graph G.

minimal cycle of a polygonal graph G: a cycle formed by the boundary edges of one of the polygons that compose G.

cycle rank γ of a graph G: the number of edges of G minus the number of vertices of G, plus 1.

degree of a vertex: the number of edges at a vertex of a graph; for directed graphs, the *outdegree* and *indegree* of a vertex are the number of directed edges out of, and into, the vertex.

dodecahedron: a polyhedron with twelve faces.

edge: a piece of curve connecting two vertices of a graph, and containing no other vertex.

cycle edge: an edge that is not a bridge.

multiple edges: if two vertices of a graph are connected by more than one edge, each such edge is called a multiple edge.

Eulerian trail: a cyclic trail that covers every edge of a graph.

face of a polygonal graph: a face of a polygonal graph G is a part of the plane bounded by a minimal cycle of G, or by the maximal cycle C_1. The face of G with C_1 as boundary is the part of the plane lying outside C_1; it is called the *infinite face*.

graph: a figure consisting of points (called *vertices*) and segments connecting some of these vertices; the connecting segments may be straight-line segments or curved segments, and are called *edges*.

bipartite graph: a graph whose vertices can be divided into two non-overlapping sets so that vertices in the same set are not connected by edges.

complete graph: a graph of n vertices with edges connecting all pairs of vertices—that is, with $\frac{1}{2}n(n-1)$ edges.

completely regular graph: a polygonal regular graph G, whose dual graph G^* is also regular.

connected graph: a graph in which every vertex is connected to every other vertex by some path.

directed graph (or **digraph**): a graph with directed edges.

dual graph G^* of a polygonal graph G: G^* is a polygonal graph, each of whose vertices corresponds to a face of G and each of whose faces corresponds to a vertex of G; two vertices in G^* are connected by an edge if the corresponding faces in G have a boundary edge in common.

Eulerian graph: a graph containing an Eulerian trail.

forest: a graph all of whose connected components are trees—that is, a graph without cycles.

Hamiltonian graph: a graph containing a Hamiltonian cycle.

mixed graph: a graph with some directed and some undirected edges.

null graph: a graph consisting only of isolated vertices—that is, a graph with no edges.

planar graph: a graph that can be drawn in the plane so that its edges intersect only in vertices of the graph.

polygonal graph: a planar graph whose edges form a polygonal net in the plane in such a way that no polygon completely surrounds another.

regular graph of degree r: if all degrees in a graph are the same, say r, then the graph is called *regular of degree* r; for directed graphs, the outdegrees and indegrees must be the same at each vertex, and be equal to each other.

tree: a connected graph without cycles.

Hamiltonian cycle: a cycle that covers all vertices of a graph.

isomorphic graphs: G_1 and G_2 are *isomorphic* if a one-to-one correspondence can be established between the vertices of G_1 and those of G_2 in such a way that pairs of vertices of G_1 are connected by an edge if and only if the corresponding pairs of vertices of G_2 are connected by an edge; for directed graphs, this correspondence must also preserve the direction of edges.

path: a route in a graph that goes through no vertex more than once.

cyclic path: a path that returns to its starting point.

polygonal net: a set of adjoining polygons in the plane, dividing the plane into polygonal pieces; the edges of these polygons are not necessarily straight lines.

polyhedron: a three-dimensional figure whose boundary consists of planes.

regular polyhedron: a polyhedron, all of whose faces are congruent polygons and at each of whose corners the same number of polygons meet.

root of a tree: any vertex that we choose to single out as starting point in a tree.

trail: a route in a graph that goes through no edge more than once.

vertex: either an endpoint of an edge, or an isolated point of a graph.

even vertex: a vertex with even degree.

isolated vertex: a vertex at which there is no edge—that is, a vertex with degree 0.

odd vertex: a vertex with odd degree.

Index